W9-BRL-091

STATISTICS BY EXAMPLE
Exploring Data

STATISTICS BY EXAMPLE
Exploring Data

Prepared and edited by the
Joint Committee on the Curriculum in Statistics and Probability
of the American Statistical Association
and the National Council of Teachers of Mathematics

Frederick Mosteller, *Chairman*
Harvard University

William H. Kruskal
The University of Chicago

Richard S. Pieters
Phillips Academy, Andover

Gerald R. Rising
State University of New York at Buffalo

Richard F. Link
Artronic Information Systems, Inc.

with the assistance of

Martha Zelinka
Weston High School, Weston, Massachusetts

ADDISON-WESLEY PUBLISHING COMPANY
Reading, Massachusetts · Menlo Park, California · London · Don Mills, Ontario

This book is in the
Addison-Wesley Innovative Series

ISBN 0-201-04873-6

GHIJK–EB–8321

Preface

Examples of statistical activity, problems, and conclusions are abundant. They appear in newspapers and magazines, in textbooks and television, and in almost every aspect of life.

Nonetheless, clear, interesting, and elementary written descriptions of such examples are hard for students and teachers to find. The reasons are various. Many examples are large in scale and require facilities for extensive handling of data. Sometimes the statistical aspects of a real-life problem appear in the middle of its solution, and when removed from this context may be hard to appreciate. The statistical part may even be invisible unless detected by an expert. Again, our tendency in mathematics teaching is to boil each problem down to an abstract skeletal minimum. Real-life statistical problems usually have extramathematical features as well as mathematical ones, and the abstraction is not ordinarily already done for the user.

In preparing the present material, the Joint Committee on the Curriculum in Statistics and Probability of the American Statistical Association and the National Council of Teachers of Mathematics thought it wise to move strongly in the direction of real-life problems and to admit the need for more explanatory text than we find in the usual mathematics book. We feel that there is a need to explain real-life problems because the student is unlikely to have had experience with any kind of statistics beyond taking averages, and the teacher with any other than the theoretical side of statistics, if that.

This series of pamphlets, *Statistics by Example*, has four parts, each intended to stand alone, but cross-referencing is given. The subtitles and reading needs of the parts are

Exploring Data	Arithmetic, rates, percentages
Weighing Chances	Notion of probability, elementary algebra
Detecting Patterns	Elementary probability, intermediate algebra
Finding Models	Elementary probability, intermediate algebra

The tables of contents and the introduction to each part give much more detail than the following brief sketches.

Exploring Data shows how to organize data tabularly and graphically to get messages and exhibit them, and it introduces elementary probability in circumstances where simple counting gets one off to a good start.

Weighing Chances develops probability methods through random numbers, simulations, and simple probability models, and presents serious analyses of complex data using informally the idea of scatter and residuals.

Detecting Patterns presents several standard statistical devices—the normal distribution, the chi-square test, and regression methods.

Finding Models encourages the student to develop models as structures for data, so that departures from the models can be recognized and new structures built.

In all four parts the keynote is real problems with real data. A few problems are treated in two separate parts, at different levels, and usually without heavy overlap.

In making cross-references to treatments in other parts, we refer to:

SBE, Subtitle, Set number.

Here *SBE* is an abbreviation of *Statistics by Example.*

We present here a body of material very different from the usual. We have, for example, not hesitated to lean occasionally upon sports in this discussion because it helps some students to learn about a statistical method in a sports context that he knows about. Then he has a chance to raise the honest "but" 's that are a part of any serious statistical analysis. On the other hand, we rarely treat well-shuffled cards or balls drawn from urns, though these are a useful stock in trade of the beginning probability student. Instead, we have tried to work with less sterilized problems.

These volumes are not mere collections of problems and examples. Rather, each represents a series of mini-learning experiences or case histories and includes exercises and projects relative to the examples.

Our material, including the exercises and projects, should help students and teachers learn to analyze statistical problems. The projects represent larger ventures that an individual or a class can carry out. By so doing, a student will learn about the practical as well as the analytic problems associated with statistical work.

Some of the sets of examples may fit well in courses other than mathematical ones—social science, biology, physics, English, economics, civics, and history.

Those who enjoy self-study may find especially congenial the approach of learning by example.

Beyond the materials we offer here, the Joint Committee has also prepared a book of descriptions of a variety of important applications entitled *Statistics: A Guide to the Unknown* (Judith Tanur *et al.*, editors, Holden-Day Publishing Co., 500 Sansome St., San Francisco, Calif., 1972). Consequently, in addition to being able to begin learning how to do statistics through examples, the student and the teacher can supplement their work by reading about applications to nearly any field of man's work and life.

How was the present book, *Statistics by Example*, prepared? The book is composed of sets of examples based upon real problems, written up by professional statisticians and experienced teachers. After they were drafted, they were sent to teachers who reviewed them for their understandability and often asked their students for furthur comments. On the basis of the criticism received, and additional critiques from members of the committee, the example sets were revised, usually more than once. The names of the teachers who formed this task force are listed below.

We owe a great debt to the teachers who provided us with so much constructive criticism. And we owe much to the authors who patiently revised their work to make it more comprehensible. The sole reward of both groups is that they have provided this new material for teaching statistics to the community.

The Committee members have served as editors for the material, and we have had the editorial assistance of Roger Carlson and Martha Zelinka.

Robert Berk, H. T. David, Janet Elashoff, Chien Pai Han, David Hoaglin, Robert Kleyle, and Perry Gluckman have helped us in a variety of ways.

We have been aided secretarially by Muriel Ferguson, Holly Grano, and Janet Quint, who also helped with the editing, as did Cleo Youtz. Holly Grano brought her expert skill to the final manuscript shown here.

Support for the Committee's work was generously provided by the Sloan Foundation.

September 1, 1972

Frederick Mosteller, Chairman
William H. Kruskal
Richard F. Link
Richard S. Pieters
Gerald R. Rising

Task Force Members

James Arnold
Thomas C. Armstrong
E. W. Averill
Crayton Bedford
Max S. Bell
James Bierden
Shirley Brady
Richard G. Brown
William G. Chinn
Jack B. Eleyet
George A. Fargo
D. Fishman
Don Fraser
Abraham M. Glass
Alice Golar
Nicholas Grant
George Grossman
Robert L. Heiny
G. Albert Higgins, Jr.
Gertrude Hillman
Donald K. Hotchkiss
John W. Hulse
Warren O. Hulser
S. Ishikawa
Ralph H. Klitz, Jr.
George C. Laumann
M. Albert Linton, Jr.
John Malak
Arthur J. Mastera
Margaret Matchett
William K. McNabb
James V. Mulloy
Christine Murphy
Fernand J. Prevost
Raymond E. Roth
Robert E. K. Rourke
James F. Sandford
L. Manlius Sargent, Jr.
Albert P. Shulte
Gerald Tebrow
Paul Tukey
Nancy C. Whitman
Martha Zelinka

Task Force Members

Introduction

This little book has behind it the idea that some knowledge of statistics is important for everybody and that some parts of statistics are easily understood by everybody. We don't mean just pie-charts and bar-graphs. Everybody runs into numbers in tables, in charts, and in graphs. People talk about averages, the average wage and the average family, about the cost of living, birth rates, and so on, but not many people know much about these concepts, because such ideas have never been widely taught in the elementary or early secondary school years.

To encourage the learning of these concepts we have brought together a number of simple examples. Each example illustrates some statistical idea and will, we hope, serve as an inspiration to the reader to develop his own similar but perhaps better examples. After all, no one knows quite as well as you just what examples will prove illuminating and exciting. And, if the statistical ideas are perhaps not as familiar to you as some other mathematical ideas, perhaps the examples will help you understand them in a fresher and more appealing manner.

Contents

Organizing and Reading Population Data

DOUGLAS H. SPICER

INTRODUCTION

Often things don't happen as expected and as a result difficult situations arise. For example, in the fall of 1971 some schools found that the enrollment in the elementary grades had gone down, while the number of students in grades 9-12 was greater than expected. Can you think of any reasons for such a change? Can you think of any difficulties it may cause? And, is it possible to predict when such changes might occur?

Look at the following statistics from the Information Please Almanac, 1971, page 685: Enrollment in Kindergarten to Grade 8 (for the United States)

	1959-60	27,602,000
	1967-68	31,642,000
	1969-70	32,600,000
Estimate for	1970-71	32,600,000

Although only four entries are given above, can these counts tell us something about the population of students in Kindergarten through eighth grade? We can see that this population increased in size from 1959 to 1970. But, why is it estimated that the 1970-71 enrollment will be no more than it was during the 1969-70 school year? To answer this question and others like it involves, in part, knowing how to read and interpret data, and how to organize it systematically. The examples in this section have been written to help develop these skills.

The author is at the Center for Population Studies, Harvard University, Cambridge, Massachusetts.

EXAMPLE 1

The birthrate of a nation is the number of births that occur during a period of time for each 1000 persons in the country. (The period of time is usually one year.) The formula used to compute birthrates is

$$R = \frac{\text{total number of births for the year}}{\text{midyear population}} \times 1000.$$

For example, if a country had a midyear population of 5000 and if there were 147 births during this year, then the birthrate for the country for the year in question was

$$R = \frac{147}{5000} \times 1000 = 29.4.$$

Birthrates for the United States for certain years from 1910 to 1970 are: 1920, 27.7; 1955, 25.0; 1910, 30.1; 1940, 19.4; 1960, 23.7; 1970, 17.6; 1945, 20.4; 1935, 18.7; 1915, 29.5; 1965, 19.4; 1925, 25.1; 1930, 21.3; 1950, 24.1.

These data are unorganized. When in this form it is difficult to answer questions about the data, and patterns that may exist are not easily seen. One way to organize the data is to construct a table. Table 1 contains the data for United States birthrates.

Table 1. Birthrates for the United States, 1910 - 1970

Year	Birthrate	Year	Birthrate
1910	30.1	1945	20.4
1915	29.5	1950	24.1
1920	27.7	1955	25.0
1925	25.1	1960	23.7
1930	21.3	1965	19.4
1935	18.7	1970	17.6
1940	19.4		

Source: Data for the years from 1910 to 1965 were obtained from the _Statistical Abstract of the United States._ The 1970 birthrate was obtained from the 1970 _World Population Data Sheet_, published by the Population Reference Bureau, Inc.

Table 1 shows that since 1955 the United States birthrate has been declining, reaching its lowest point in 1970.

Exercises for Example 1

1. The 1970 birthrate is the lowest in the table. In which year did the next lowest birthrate occur? What crisis was taking place at this time?

2. Some people feel that the United States birthrate will probably continue the decline begun in 1955. Does past history support this claim? Describe in words how the United States birthrate behaved from 1910 to 1970.

3. Another important rate to know when studying population matters is the death rate. The death rate is similar to the birthrate except that it gives the number of deaths for each 1000 people over a one-year period of time. Construct Table 2 using the following data for the death rate for the United States: 1920, 13.0; 1965, 9.4; 1940, 10.8; 1955, 9.3; 1915, 13.2; 1930, 11.3; 1950, 9.6; 1970, 9.6; 1945, 10.6; 1935, 10.9; 1910, 14.7; 1925, 11.7; 1960, 9.4. [You need Table 2 later.]

EXAMPLE 2

Knowing how fast its population is growing helps a country to plan for future educational, recreational, and other kinds of needs. In order to determine how fast the population is growing, one must know not only the birthrate and the death rate, but the rate of net migration as well. The rate of net migration takes into account the number of people entering a country (immigration) and the number leaving (emigration). The rate of net migration will be positive if the amount of immigration is greater than the amount of emigration. It will be negative if the amount of emigration is greater than the amount of immigration. Table 3 contains rates of net migration for the United States.

Table 3. Rates of net migration for the United States, 1910 - 1970

Year	Rate of net migration	Year	Rate of net migration	Year	Rate of net migration
1910	10.4	1935	0.4	1955	1.5
1915	5.7	1940	0.4	1960	1.5
1920	5.7	1945	0.7	1965	1.5
1925	3.5	1950	0.7	1970	2.0
1930	3.5				

Source: Data for the years from 1910 to 1965 were obtained from the Statistical Abstract of the United States. The 1970 rate was computed using data from the 1970 World Population Data Sheet, published by the Population Reference Bureau, Inc.

We now have enough information to compute the population growth rate of the United States for the years appearing in Tables 1, 2, and 3. Let's compute the rate for 1955 to illustrate how it is done.

First determine the value of the expression: birthrate - death rate + rate of net migration. For 1955 the value is: 25.0 - 9.3 + 1.5 = 17.2. This means that for every 1000 people in the United States in 1955, 17.2 people were added to the population. However, population growth rates are usually expressed as percentages. Thus we next divide 17.2 by 10, obtaining an annual population growth rate for 1955 of 1.72%.

Exercises for Example 2

1. Compute the annual rate of population growth for the United States for the years appearing in Tables 1, 2, and 3. Construct Table 4 using these data. [Table 2 was constructed for Exercise 3 above.]

2. Describe in words the pattern displayed by the data in Table 4.

EXAMPLE 3

When you are interested in displaying a trend or pattern formed by population data, a graph may be more useful than a table. The data contained in Table 4 are shown in the line graph in Fig. 1.

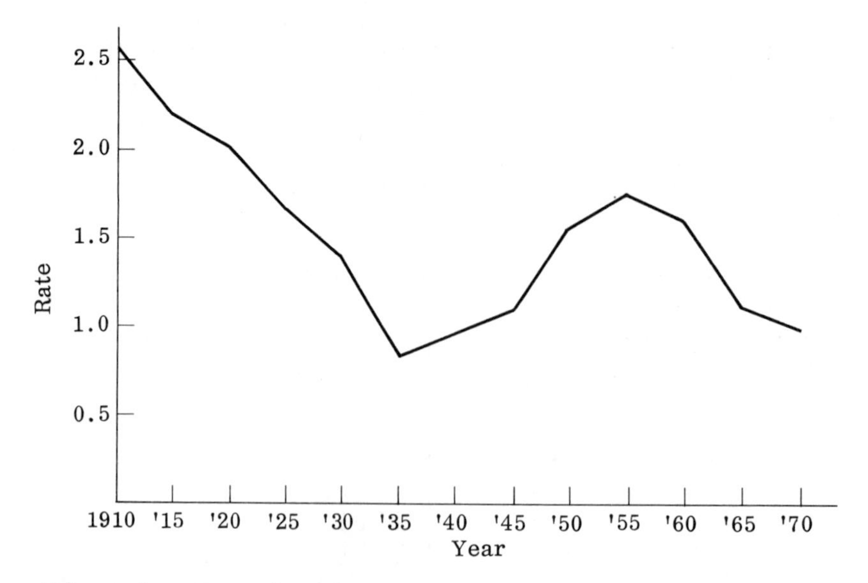

Fig. 1. Population growth rate of the United States, 1910 - 1970.

As the graph shows, the rate of population growth of the United States seems to be declining again after having increased from a low point reached in 1935.

Exercises for Example 3

1. Using the graph in Fig. 1, is it possible to determine in which year the most and in which year the fewest number of people were added to the United States population? If so, in which years were the most people added? the fewest? (You might go to the Statistical Abstract of the United States to obtain an estimate of how many people were actually added during the years in question.) If you cannot tell using the graph, explain why not.
2. Construct line graphs for each of Tables 1, 2, and 3. Which of the three variables--birthrate, death rate, or rate of net migration--appears to have had the greatest influence on determining the United States rate of population growth?

EXAMPLE 4

Because the United States rate of population growth has always been greater than zero the population of the country has continually increased. As the population has increased, a larger percentage of the people have begun living in and around cities. This is called urbanization. The trend toward urban living is illustrated by the bar graph in Fig. 2.

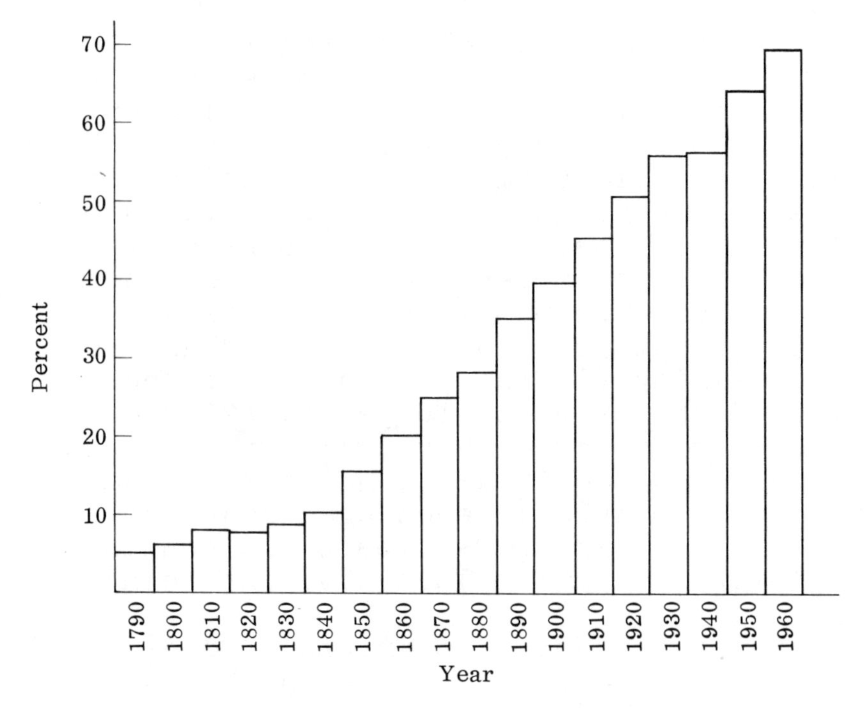

Fig. 2. Percentage of U.S. population living in urban areas, 1790-1960. Data were obtained from the Information Please Almanac, 1968.

The graph shows that the percentage of the population living in urban areas increased from just over 5% in 1790 to almost 70% in 1960.

Exercises for Example 4

1. Does the increase in the percentage of people living in urban areas mean that fewer people lived in the country in 1960 than in 1790? Use an appropriate reference book to check if the number of people living in the country decreased as the percentage of the people living in cities increased.

2. What are some reasons why a greater percentage of the United States population is living in and around cities?

3. Although more and more people are living in urban areas, many of the cities with the most people are losing population. Using data from the Statistical Abstract of the United States or from a similar book, make a bar graph that compares the changes in population from 1950 to 1960 of the 10 largest cities in the United States. Which cities lost population from 1950 to 1960? Why do you think these cities are losing population?

Suggested project

Assume that the students in the class, together with their parents and their brothers and sisters who live at home, form the population of a hypothetical country. (The students can choose a name for the country.) Have each student make a table containing information about the number of people in his or her family for each of the previous 10 years. For each year also have students record the number of males and females in the family, and the age of each member of the family.

Pool the data the students have collected and construct tables containing (1) the population of the country for each of the previous 10 years; (2) the birthrate of the country for each year; (3) the number of males and females in the population for each year; and (4) the number of people in each age group for each year, beginning with the group 0-4 years and continuing up in 5-year intervals with the last division being 65 years and over.

For (1) and (2) construct line graphs. Combine (3) and (4) and construct a special kind of bar graph called a population pyramid. Examples of population pyramids can be found in U. S. Census Reports.

The tables and graphs for the population of the hypothetical country can be compared to tables and graphs presenting similar information about the United

States or other real countries. If other classes are available, then comparisons can also be made among different hypothetical countries.

Fractions on Closing Stock Market Prices

FREDERICK MOSTELLER

INTRODUCTION

Often numbers appear in groups. For instance, if you measure the height of each of 30 students, there may be 5 who are 62 in. high, 6 who are 63 in., 2 who are 64 in., 8 who are 65 in., 4 who are 66 in., and 5 who are 67 in. When we have such a grouped set of measurements, we talk about a frequency distribution, since the height 62 appears five times or "with frequency 5". Since such frequency distributions occur in many circumstances, you need to have some experience in gathering data about them and in continuing to gather until you have enough.

What does enough mean? Usually what we want is to develop a theory which involves the data, and we want enough to support the theory or to disprove it. We want to discover whether the data contain regularities or quirks that larger samples will remove. Usually, the more specific we can be about the regularity, the better. These examples offer you an economical way to get experience of this kind by looking at some stock market prices in a daily newspaper. There are also a few exercises for practice. You won't learn how to make a killing in the market, but you will learn how to begin thinking about the ways samples may differ, how to look for regularities, and how to try to verify them.

The author is at Harvard University, Cambridge, Massachusetts.

EXAMPLE 1. FIRST INVESTIGATION

Obtain the closing New York stock market prices from a newspaper that carries fairly complete information. Be sure to get the closing rather than the noon or the late prices. Things may not be the same during the day as they are at the close--the stock specialist, who handles a particular stock, has some options open at the end of the day.

The prices are recorded to 1/8ths. For example, Anacon 25½. This means Anaconda Copper closed at 25½ dollars a share. On one occasion (May 14, 1970) the whole entry for Anaconda read

Year High	Year Low		Divi- dend	Hundreds sold	1st	High	Low	Close	Change
$32\frac{1}{4}$	$25\frac{1}{2}$	Anacon	1.90	458	$25\frac{3}{4}$	$25\frac{7}{8}$	$24\frac{7}{8}$	$25\frac{1}{2}$	$-\frac{1}{2}$

This line says that the previous high and low prices for the present year were \$32¼ and \$25½, that the stock was Anaconda, that it payed a dividend of \$1.90 per year, that it sold 458 hundred shares (45,800 shares) that day, that its opening price was $\$25\frac{3}{4}$, that the day's high and low were $\$25\frac{7}{8}$ and $\$24\frac{7}{8}$, that it closed at \$25½, and gained \$-½, that is, it lost \$½ from the May 13 close.

One might suppose that the final eighths would be equally likely since prices move by 1/8ths. Take a sample of stocks, obtain the fraction on the closing prices (in our example, ½), convert to eighths and get the number of times each number of eighths occurs; this is called a frequency distribution and is illustrated in Table 1. How does your table compare with Table 1? In making your frequency distribution it would be well to make more than one set of the same size, perhaps 3 of 50 each or 3 of 100 each, so as to get some notion of the repeatability of your finding. We take 50 or 100 so as to get a reasonable average of the number of occurrences of each eighth, about 6 to 12 measurements for each. Examine the frequency distribution and discuss the results. Look for special properties of the frequencies. After you get your finding, do at least one more sample to see if you find it still gives the same general results. (You and your classmates should gather data from different days so that these can be pooled.)

Table 1. Frequency distribution of final fraction on closing prices of the New York Stock Exchange

Fraction	No. of 8ths	Set 1 No.	Set 1 %	Set 2 No.	Set 2 %	Set 3 No.	Set 3 %	Total No.	Total %	Replications First	Replications Second
0	0	14	23	18	30	10	17	42	23	19	21
1/8	1	3	5	4	7	4	7	11	6	8	6
1/4	2	6	10	8	13	7	12	21	12	10	11
3/8	3	2	3	2	3	7	12	11	6	12	6
1/2	4	16	27	9	15	13	22	38	21	11	26
5/8	5	5	8	8	13	4	7	17	9	16	7
3/4	6	7	12	8	13	11	18	26	14	18	16
7/8	7	7	12	3	5	4	7	14	8	6	7
Totals		60	100	60	99*	60	102*	180	99*	100	100

Set 1: Abbott Lab to Am Enka
Set 2: Am Exp to Atchison
Set 3: Cur Pub to Dyna Am
First replication: Eagle P to Gemini Cap
Second replication: Mac Avd F to Nat Airlin

*Percentages may not add to 100 because of rounding.

Solution. The New York Times, February 7, 1968, gave the 3 sets of 60 ending fractions for closing stock prices of February 6 shown in Table 1. (The student at first chose a unit of 60 instead of 50 or 100, rather an inconvenient size. Why is 60 inconvenient?)

Observe that the big frequencies are associated with 0, 1/2, 3/4, 1/4 and that the odd eighths are relatively rarer. The even frequencies add up to 127 versus 53 for the odds, or 70% even eighths. It looks as if there is something about the market that leads to there being many more even eighths than odd eighths. Now we have a hypothesis, not only that the fractions are not equally likely, but more specifically that the even ones are more frequent, and we need to check it on fresh data. Let us take an additional hundred and see how close to 70% we get for the even eighths.

Our new results are not as compelling as the earlier observations, but did give 58% even eighths. This makes us consider whether an additional count would be worth having. A second set of 100 supported the original finding very well, and its percentage of even eighths is 74. All told, then, we think that even eighths are more than twice as likely as odd eighths in closing stock market prices.

Before reading further can you think of any reason why the even eighths should be so much more likely than the odd eighths?

Exercises for Example 1

1. Check that 70% are even eighths in the first three sets of Table 1.
2. Check how close the entries in today's market come to 70% for even eighths.
3. In Table 1, check to see whether the 4 largest fractions get more than half the frequencies.
4. Do any fractions stand out as especially frequent?

EXAMPLE 2. SECOND INVESTIGATION

Upon reflection, some students have suggested that high-priced stocks may move by large units rather than small ones. They suggest that if we restricted our data to stocks selling under $10 per share we might find that all eighths occurred with about equal frequency. Gather new data and test this idea.

Solution. Table 2 gives the frequency distribution of closing fractions for stocks closing at 10 or under Friday, November 21, 1969, (Source: Boston Herald Traveler, November 23, 1969, p. 48) for the American Stock Exchange. We couldn't use the New York Stock Exchange because too few stocks sold below 10. The table shows that even eighths occur about 55% of the time. The average frequency is about 40 ($317/8 \approx 40$) and so we see that the counts are not close to being equal. The table shows that 0 is very frequent, 1, 2, 3 eighths are infrequent, 4, 5 are more frequent than average, and 6 and 7 are about average. Thus we do not find the "saw-toothed" distribution we observed in Table 1 for all stocks. But we still do not find a flat distribution. We need some objective way of deciding when a frequency distribution is close to a theoretical one and when it is not. This requires a statistical method treated in reference [1] and applied to this problem in reference [2].

Table 2. Frequency distribution of final fraction on stocks selling at or below 10 on the American Stock Exchange

Fraction	Eighths	Frequency	%
0	0	60	19
1/8	1	30	9
1/4	2	29	9
3/8	3	27	9
1/2	4	47	15
5/8	5	49	15
3/4	6	37	12
7/8	7	38	12
		317	100

Exercises for Example 2

1. Graph the data of Table 1 and that of Table 2 side by side. Explain "saw-toothed" as applied to Table 1.

2. Make a table for closing fractions for stocks closing at 50 or more and one for those closing at 100 or more. Do these tables confirm or deny the suggestion that higher-priced stocks move by larger units than low-priced ones?

3. Do the data in Table 2 suggest that the largest four fractions have more than half the frequencies?

4. Go to the World Almanac or similar book, or to a physical and chemical handbook, and collect a frequency distribution of the "first significant digits" (1,2,3,4,5,6,7,8,9) for some type of data--for instance, voting statistics, heights of mountains, physical measures--and discuss the outcome. [In all the following numbers, the digit 2 is the first significant digit: 2897000, 2897, 2.897, 0.2897, 0.0002897.]

5. Guess what proportion of major league batting averages have a first significant digit of 2. Try to find published data to check on your guess.

6. Find out if noon stock prices have distributions of final fraction that are like those of closing prices for stocks over 50.

References

[1] SBE, Weighing Chances, Set 6.

[2] SBE, Weighing Chances, Set 7.

References labeled SBE are to one of the four parts of Statistics by Example.

Characteristics of Families and Their Members

JOSEPH I. NAUS

INTRODUCTION

These four examples discuss various kinds of information presented in a table of data about 25 different families. Some of this information involves specific quantities, and some of it general qualities associated with the families; both the quantities measured and the qualities counted are called variables. The first example illustrates the different kinds of variables. It will show you that sometimes the same variable may be viewed in several different ways. The second example describes how a frequency distribution can be used to summarize information about a variable. It also shows how modifications of this approach can be used, depending on the nature of the variable involved. You will see how this approach also points out features of the data for further study. The third example illustrates another device used to summarize information, the average value. But it also demonstrates the importance of understanding the properties of averages in order to know what they can and cannot do.

Usually we can get more information looking at two variables together than we can by looking at each of them separately. In the fourth example you will see several ways to do this.

The author is at Rutgers University, New Brunswick, New Jersey.

Sociologists study human groups and how they interact. The following examples are based on a recent study by a sociologist in a small city. Studies of this kind are made for reasons such as to find how many children there are, how many elderly or unemployed people are eligible for welfare, or how many people want to attend an evening high school or community college. More often, we want to know how these variables relate to one another. Studying the people living in every house may be too expensive or take too long; so the sociologist picks out certain houses and interviews only their residents, usually in such a way that they will be approximately like the others, so he can make judgments about all the people in the city from his study of a relatively small group.

How he selects the houses he visits is an interesting and important question. It is a problem in *sampling*, and we will discuss it more in Example 4. Obviously, the results of the study will depend heavily on which houses are picked for the interviews. But forgetting this for awhile, let us see how the sociologist goes about making his study after he has decided which houses to visit. The first house is that of Mr. Joseph Green. (The data used are real, but the names are fictional.) He asks Mr. Green some questions and records the answers.

Case study 1. Mr. Joseph Green is a 70-year-old male. He is married and has three children. Both Mr. Green and his wife went to high school, but neither graduated.

At the next house, the sociologist interviews Mrs. John Smith and writes her case study.

Case study 2. Mrs. John Smith is a 28-year-old female. She is separated from her husband and has three children. She went to high school, but did not graduate; her husband is a high-school graduate.

At each of 25 houses he interviews the person and writes up the answers as case studies, which are summarized in Table 1. The following examples are based on these data.

Table 1. Summary of information from 25 case studies

Case study	Age	Sex	Marital status	Number of children	Educational level Husband	Wife
1	70	male	married	3	some high school	some high school
2	28	female	separated	3	high-school grad.	some high school
3	47	male	married	1	some high school	high-school grad.
4	48	male	married	3	high-school grad.	some high school
5	23	male	married	0	college grad.	high-school grad.
6	69	female	divorced	3	--	grad. eighth grade
7	31	female	married	1	some grad. work	Master's degree
8	70	female	single	0	--	high-school grad.
9	80	female	widowed	0	--	less than 8 grades
10	37	female	married	3	high-school grad.	some high school
11	65	female	single	0	--	some high school
12	71	female	widowed	2	--	some high school
13	41	female	separated	3	high-school grad.	high-school grad.
14	70	male	married	5	some high school	less than 8 grades
15	56	female	married	1	grad. eighth grade	grad. eighth grade
16	34	female	married	4	some high school	grad. eighth grade
17	48	female	married	6	high-school grad.	some high school
18	43	female	divorced	3	--	some high school
19	50	female	married	1	high-school grad.	some high school
20	24	female	married	2	high-school grad.	high-school grad.
21	23	male	married	1	college grad.	high-school grad.
22	47	female	married	4	Ph.D. degree	Ph.D. degree
23	63	male	single	0	high-school grad.	--
24	31	female	married	4	high-school grad.	some high school
25	21	female	separated	3	some high school	some high school

EXAMPLE 1. TYPES OF OBSERVATIONS

The first two rows of Table 1 describe Mr. Joseph Green and Mrs. John Smith. Each row in Table 1 describes an individual. The sociologist also seeks to describe the group. Each column in Table 1 gives the measure of a characteristic of the group. The column headed "Age" lists the ages of the 25 people interviewed.

The variable "Age" takes values that are numbers; the variable "Sex" takes values that are qualities (being male or female). We call age a quantitative variable and sex a qualitative variable.

Age and Number of Children are both quantitative variables, but they differ in an important way. We can measure exactly how many living children a person has. A person can have zero, one, two, three or more children, but a fractional number of children is not possible. Rather, there is a discrete set of possibilities for number of children; exactly zero, one, two, three or more children. We say that Number of Children is a discrete variable.

We cannot measure a person's age exactly; we can only specify it within an interval (say, to the nearest minute). Though recorded here in whole years, the person's age can fall anywhere in the continuous range of this yearly interval. We call Age a continuous variable, even though the measuring we do is discrete.

Exercises for Example 1

1. Is Marital Status a quantitative or qualitative variable?

2. The sociologist chose to record Educational Level of husband and wife in terms of categories of the form "some high school", "high-school graduate", "some college", and so on. He felt that there was a qualitative difference between "some high school" and "high-school graduate". Two individuals with some high school, but not graduates, are both in the category "some high school". They both possess this same quality. The sociologist chose to view Educational Level as a qualitative variable.

   a) Even though Educational Level is regarded here as a qualitative variable, it differs in one important respect from the Marital Status variable. Can you see what it is? Is there not a natural order connected with the Educational Level variable which is missing in the other? Put the various categories in this column in an order which seems natural to you.

   b) The sociologist could record the years and fractions of years of schooling. Would he then be viewing the underlying variable as quantitative or qualitative, continuous or discrete?

   c) The sociologist could record the number of courses successfully completed; is the underlying variable quantitative or qualitative, continuous or discrete?

EXAMPLE 2. FREQUENCY DISTRIBUTIONS AS A SUMMARY OF DATA (ONE-OUTCOME VARIABLE)

The sociologist presents 25 case studies. He can greatly condense the information in the columns of Table 1 to give a compact description of the group. There are 18 females and 7 males. He can summarize these data in the frequency distribution in Table 2.

Table 2. Frequency distribution of sex of 25 individuals interviewed

(1) Sex	(2) Frequency	(3) Relative frequency
Male	7	0.28
Female	18	0.72
Total	25	1.00

Ordinarily, the first column in a frequency distribution is headed by the name of the variable and consists of the possible values of the variable. The second column gives the frequency or number of individuals who take on the value of the variable shown at the left. The third column gives for each value of the variable the relative frequency (fraction or proportion) of individuals who take on that value of the variable.

The frequency distribution for Sex shows that a large fraction of the 25 individuals interviewed are female. Although about one-half of all individuals in the general population are male, of these 25 individuals, only 28% are males. Is this city different from the general population? Or was the sociologist unlucky in his choice of the 25 individuals? Or did his method of picking the individuals lead to a large fraction of females? These questions led the writer to find out that the survey was a household survey conducted during the day, so that the data are unusual in certain ways. Summarizing the data in frequency distributions focuses attention on important questions raised by the whole body of the study.

For the variable Number of Children we see that there are 5 families with no children, 5 families with 1 child each, 2 families with 2 children each, 8 families with 3 children each, 3 families with 4 children each, 1 family with 5 children, 1 family with 6 children. Table 3 summarizes the distribution of the variable Number of Children. Column (1) is headed by the name of the variable (Number of Children), and consists of the values of the variable for the 25 families. Column (2) lists the frequency of families with the number of children given by the corresponding entry in column (1). Column (3) gives the frequency relative to the total number of families.

Table 3. Number of children per family for the 25 families

(1) Number of children	(2) Frequency (number of families)	(3) Relative frequency
0	5	0.20
1	5	0.20
2	2	0.08
3	8	0.32
4	3	0.12
5	1	0.04
6	1	0.04
Total	25	1.00

The frequency distributions in Tables 2 and 3 have condensed the columns for Sex, and Number of Children, which are both discrete. We can list the possible values that a discrete variable can take. How do we list the values for a continuous variable? or for one with many values, like age in years?

We can construct a frequency distribution listing the different ages that appear in the sample, but the frequency distribution will list 19 different values (out of 25 ages). This does not achieve much condensation of the original data. We can achieve better condensation by rounding the ages to the nearest ten years. We round 71 to 70, 37 to 40. How should we round 65? There are several approaches. We can flip a coin (but this makes it hard to check someone else's work). We can round to the nearest even number (65 is rounded down to 60, 75 is rounded up to 80). In the present example we round up, because a person who reports his age is 65 is usually over 65. Table 4 gives the frequency distribution of the rounded ages.

Table 4. Rounded ages of the 25 people interviewed

(1) Rounded age	(2) Frequency (number of interviewed that age)
20	4
30	4
40	3
50	5
60	2
70	6
80	1
Total	25

A convenient way to set up the frequency distribution in Table 4 is shown in Table 5. We write down next to each rounded age the ages in Table 1 that will be rounded to those values. For example, all ages in Table 1 from 55 through 64 will be rounded down or up to 60. We say that the <u>stated class interval</u> corresponding to the rounded age 60 is 55-64. Table 5 lists the stated class intervals.

Table 5. Rounded ages of the 25 people in the study

(1) Stated class interval	(2) True class interval	(3) Rounded ages	(4) Frequency
15-24	15.0-24.9+	20	4
25-34	25.0-34.9+	30	4
35-44	35.0-44.9+	40	3
45-54	45.0-54.9+	50	5
55-64	55.0-64.9+	60	2
65-74	65.0-74.9+	70	6
75-84	75.0-84.9+	80	1
Total			25

We can also write down the true ages that correspond to the rounded age 60. A person listed in Table 1 as 64 could be 64 years 11 months and 29 days. The true ages that correspond to an age rounded to 60 could be any age between 55 and slightly under 65. We say that the <u>true class interval</u> corresponding to the rounded age 60 is the interval 55 - 64.9+. The true class intervals are also listed in Table 5.

The two types of class intervals serve different purposes. The stated class intervals simplify the classification of the ages in Table 1 into the categories in Table 4. The true class intervals tell us the range of possible ages for the individual in the interval.

Exercises for Example 2

1. Construct the frequency distribution for Marital Status. Show both frequency and relative frequency.
2. In Table 1 round the ages to the nearest 20 years. (That is round to one of the numbers 0, 20, 40, 60, or 80.) Show stated class intervals, true class intervals, and frequency distribution.
3. Suppose that the sociologist, when asking people their ages, asked them for their birthdates and then rounded the ages to the nearest year. A person with recorded age 65 would then be between 64 years 6 months and 65 years 6 months. Suppose we are rounding ages to the nearest 10 years. How should we then round the age 65?

EXAMPLE 3. AVERAGES AS SUMMARY QUANTITIES FOR DISTRIBUTIONS, MEASURES OF LOCATION

The previous section shows how to summarize information about variables into a frequency distribution. For example, all the information about the individual variable Number of Children is summarized in Table 3. The sociologist seeks to condense the information about this variable even further. He seeks a typical or average measure of the number of children per family. He might observe that there are more families that have 3 children than any other number of children. The most frequent, or modal, number of children gives some indication of the typical number of children.

The sociologist might also observe that half the families have 3 or fewer children, and half have 3 or more children. This number is called the median number of children. In our case there is a family with 3 children, and we might call it the middle family in size. But suppose in another study of 24 families we found that the frequency distribution was

Number of children	Frequency
0	1
1	4
2	7
3	0
4	8
5	4

The median number of children is again 3, but there is no middle family since no family has exactly 3 children. Even so, the median gives another indication of the typical number of children in the families of our survey.

The sociologist might note that there are 56 children in the 25 families, or an average of 2.24 children per family. This average is called the arithmetic mean (or, for short, just mean) number of children per family. Notice again that, just as with the median, there may not be--in fact usually is not--any family with the mean number of children.

The mean, median, and mode are averages, all measures of the position of the center of a frequency distribution; the sociologist might choose any one to describe the situation in the group of 25 families. His choice depends on the expected use for the average. The mean, median, and mode have many properties, not all of which are the same. Many misuses of statistics arise from assuming an average has a property that in fact it does not possess. Let us look at an example where such misuse might occur.

Illustration. A city manager seeks to estimate the total cost of a recreational program. The program costs $10 per child and is to be instituted in a community of 1000 participating families. He reads in a study done on the community that the average family has 3 children. He calculates that since there are 1000 families, and the average family has 3 children, there are 3000 children. Since the program costs $10 per child, he calculates the cost of the program as $30,000.

But his calculations are based on the assumption that, if you know the average number of children per family and multiply by the total number of families, you will find the total number of children. The report that stated that the average family has 3 children might have used the mean, the median, or the mode as its average. The manager assumed by his calculations that all averages share the property

$$\text{average number of children} = \frac{\text{total number of children}}{\text{number of families}}.$$

This equation defines not the median or the mode, but the arithmetic mean. Suppose that in this community the median number of children is 3, but the mean number of children is 2.24. The program would actually cost $22,400 instead of the estimated $30,000.

This illustration points out that the use to be made of the average should determine the average to be chosen. Sometimes, but not always, the uses will be clearly related to the definition of the average, so that the choice of the average is straightforward. The following exercise considers the use of the median as an average, as well as the mode and the mean.

Exercises for Example 3

1. In a 1965 study of a population of families, a sociologist found that the average family had 3 children. In a 1970 study of the same population of families, he found that the average family had 3 children. He concludes that the typical family in 1965 had the same number of children as the typical family in 1970.

   Another sociologist studying the same population asked each family in 1970 how many additional children it had between 1965 and 1970. He found that the average family increased by 1 child over the period. The two sociologists were surprised at the contradictory results as both of them had used the same average, the median.

a) Should the median number of children in 1965, plus the median increase between 1965 and 1970, equal the median number of children in 1970? To answer this question consider the example for the 25 families listed in Table 6. Compute the median for each of the three columns in Table 6.

b) Suppose the sociologists had both used the mode as their average. Could they still have had the contradiction? (Check in Table 6.)

c) Could the sociologists have had a contradiction if they had used the mean? Assume that the population consists of the same set of families. Use Table 6 to illustrate your reasoning.

Table 6. Number of children per family for 25 families

Case study	Number of children 1965	Increase in children 1965-1970	Number of children 1970
1	3	0	3
2	3	2	5
3	1	1	2
4	3	1	4
5	0	3	3
6	3	0	3
7	1	2	3
8	0	0	0
9	0	0	0
10	3	1	4
11	0	0	0
12	2	0	2
13	3	1	4
14	5	0	5
15	1	0	1
16	4	1	5
17	6	1	7
18	3	1	4
19	1	0	1
20	2	1	3
21	1	2	3
22	4	1	5
23	0	0	0
24	4	1	5
25	3	1	4
Totals	56	20	76

EXAMPLE 4. ANALYSIS OF RELATED DATA (MORE THAN ONE-OUTCOME VARIABLE)

The sociologist has measured several variables on 25 families, to form case studies; the 25 case studies are summarized in the rows of Table 1. Example 2 illustrates how to summarize information in the columns of Table 1 to study one variable at a time. But suppose the sociologist is primarily interested in the relation between two or more variables. The present example illustrates how to cross-tabulate the data to study the relationships between several variables.

For instance, the sociologist has studied the distributions of Number of Children and Marital Status for the group. He might then seek to describe the relationship between Number of Children and Marital Status. He can summarize this information for the group by constructing Table 7.

Table 7. Cross tabulation of marital status versus number of children

	Marital status					
Number of children	Married	Single	Separated	Divorced	Widowed	Total
0	1	3	0	0	1	5
1	5	0	0	0	0	5
2	1	0	0	0	1	2
3	3	0	3	2	0	8
4	3	0	0	0	0	3
5	1	0	0	0	0	1
6	1	0	0	0	0	1
	15	3	3	2	2	25

The rows in Table 7 are headed by the values of the variable Number of Children. The columns in Table 7 are headed by the possible values for Marital Status. The entries in the table are the frequencies that a particular Marital Status--Number of Children combination appears in the group. For example, the frequency 5 under the column headed Married, and across from Number of Children 1, means that 5 of the individuals interviewed are married with exactly 1 child. The entries in the table give the frequency that combinations of values for the variables appear jointly, and are referred to as the joint distribution. The two-way classification lets us study relationships between two variables, Marital Status and Number of Children. For example, from the summary in Table 7 we see that all 5 separated and divorced individuals interviewed have exactly 3 children. The sociologist seeks to explore and explain such observed relationships. For the present example, there are several possible explanations for the observed relationships.

1. There is something about having exactly 3 children that increases the chance of a person getting separated or divorced. Perhaps at 3 children the stresses on a marriage become very great, and if the marriage survives this, then it is likely to continue to survive.
2. A person who gets separated or divorced usually does so in the early part of marriage, and so does not have the opportunity to have more than 3 children.
3. The sample is small, consisting of only 5 separated and divorced individuals. Perhaps if we took a bigger sample we would see a more even spread of number of children among separated and divorced.
4. The sample was conducted by interviewing individuals at home during the day. A separated or divorced woman with no children would very likely be out working. Even women with 1 or 2 children would more likely be out working than women with 3 children.

Explanation 1 suggests that having 3 children "causes" a higher chance of divorce. Explanation 2 states that divorce causes a limitation of the number of children. Explanations 3 and 4 state that the observed relationship might not carry over for the general population, but may be due to chance situations in a small sampling, or the results of a biased method of sampling. Statistical theory can eliminate explanation 3; proper sampling will indicate whether or not explanation 4 is valid. If proper sampling still reveals the same relationship between marital status and number of children, the sociologist might conduct detailed interviews to find out whether or not the stresses of 3 children lead to an unusually large chance of divorce.

Exercises for Example 4

1. What is the interpretation of the 3 in the cell under the column "Single", and across from the row for zero Number of Children?
2. Construct a cross-tabulation of Educational Level of Husband against Educational Level of Wife from Table 1. What is an interpretation of most of the frequencies falling in or near the diagonal of the table?
3. Construct a cross-classification of Age (rounded to nearest 10 years) and Educational Level (use three levels: some high school or less, high-school graduate, and education past high school).

Points and Fouls in Basketball

ALBERT P. SHULTE

INTRODUCTION

Sometimes you may think two different sets of numbers are related in some way, though not perfectly. How can you decide if they are related or not? If they are related, is there any reason why they should be, or is it purely accidental? Does a change in one of the variables cause a change in the other? This example is designed to help answer the first question, but usually the second and third ones are much harder, and people often disagree about the answers. One purpose of this example is to show that some questions, even in the mathematical sciences, do not have hard and fast, or certain, answers. Furthermore, a superficially observed relationship may not properly suggest the cause.

EXAMPLE

The following data come from the records of the junior-varsity basketball team at Waterford-Kettering High School, Waterford Township Schools, Michigan, in two separate years. The two sets of players are entirely distinct in the two years.

The author is with the Oakland Schools, Pontiac, Michigan.

Table 1. First year

Player	Total points scored	Personal fouls committed
Bogert	1	0
Bone	0	0
Campbell	0	0
Forbes	2	0
Godoshian	6	6
Graham	48	17
Madill	21	7
Manning	75	31
McGrath	18	9
Nutter	2	4
Nyberg	42	24
Shipman	60	15
Spencer	37	16
Watson	3	0

Table 2. Second year

Player	Total points scored	Personal fouls committed
Brummett	2	1
Cooper	75	24
Felice	0	1
Hook	59	18
Hurd	9	9
Kampsen	7	3
McPartlin	35	5
Pointer	46	20
Schuback	0	1
Wilson	2	3
Zuelch	57	22

Even a quick look at the tables makes one feel that a strong relationship exists between the number of points a basketball player makes in a season of play and the number of personal fouls he commits.

In Fig. 1, the data for the first year as given in Table 1 have been plotted, with personal fouls committed on the horizontal axis and total points scored on the vertical axis.

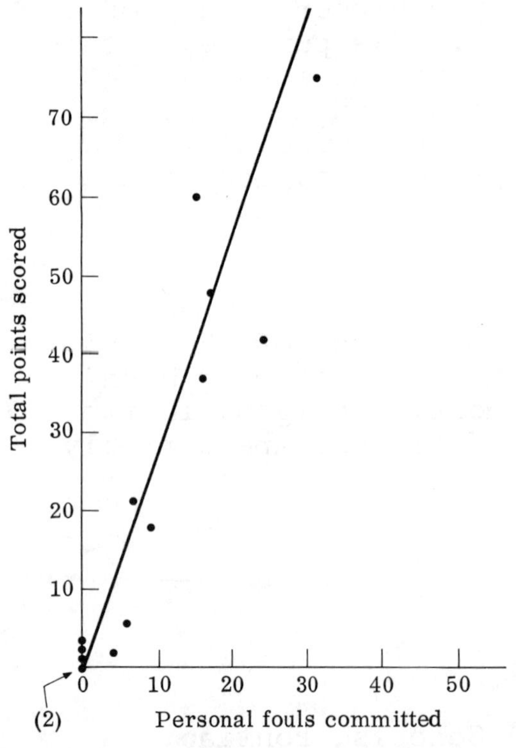

Fig. 1. Graph of data from Table 1 together with line through the origin fitted by eye.

Exercises

1. The simplest kind of relation is a straight line through the origin. Draw a straight line through the origin that "fits" the data points, that is, draw a line the data points lie on or near. Of course, there is no one line that everyone will choose, but if you use a transparent ruler or a black thread, you should be able to pick out a line with a pretty good fit. Some points may lie on it, some above it and some below it, but what you should try to get is a line which shows the general trend. We have drawn in one such line. Maybe you can draw one which you feel fits the data better. In that case, use it for the following exercises. Otherwise, use the one already drawn.

2. (For those who know some algebra.) Write the equation of the fitted line in $y = mx$ form, where y is the total points scored and x is the number of personal fouls committed.

3. Use the line to predict the number of total points expected for a given number x of personal fouls. Do this for $x = 7, 9, 15, 17$. How closely do the results approximate the corresponding values in the table? Do it the other way; that is, suppose a player scored 5, 10, 20, 30, or 40 points. How many fouls does the line indicate he might be expected to have made? What percentage of the points gives approximately the number of fouls?

Now let's think a bit more about the relationship that seems to exist between the number of points a basketball player scores in a season and the number of personal fouls he commits. The table and the graph both indicate that the more fouls he commits the more points he scores. On the graph this is indicated by the fact that our fitted line goes up to the right. But does this make sense? One might think that if a player commits a lot of fouls he would be thrown out of the game and lose his chance to score many points. Surely no coach would coach a player to make lots of fouls in the hope that he would therefore score more points. The crucial word in the previous sentence is "therefore". Is the fact that a player commits more fouls the reason that he scores more points? Of course not. But maybe both of these acts, more fouls and more points, are the results of some other act. Perhaps they are both due to some players playing more of the total game time than others. Or maybe both are due to some players being more aggressive than others, and that the more aggressive player both scores more points and commits more fouls.

When two sets of numbers are related as these are, we say that they have a high positive correlation. But just because there is a high correlation, as between fouls and points, we cannot assume that one is the cause of the other. If all the players on a team were of nearly equal ability, then the correlation between the time they played and the points they scored would be high, and we would be nearly sure that if one player scored many more points than another, it would be because he played many more minutes.

A graph such as that in Fig. 1 can show that a high degree of correlation exists, but it does not tell us why. Why may be hard to see and harder yet to verify. For example, the population of the United States and the annual consumption of liquor in England have both increased steadily in the last fifty years, so that there is a high correlation between them, but neither one caused the other. The vague common cause is the industrial revolution. As populations rise, we make more of the products people want. And so populations and production statistics are likely to increase together. Similarly, variables related to business fluctuations are likely to vary together.

Comparing diagrams like that in Fig. 1 with others, like the one we ask you in a later exercise to draw from Table 2, is often difficult by eye. Therefore, we want a number that can be calculated for each case to make the comparison easier.

We want a standard way to get the slope. If we regard a point as having coordinates (x,y), then the slope of the line from the origin to the point is y/x. If we have several points, one choice for the slope of the line is

$$m = \frac{\text{sum of } y\text{'s}}{\text{sum of } x\text{'s}} .$$

For Table 1 we get the slope

$$m = \frac{315}{129} = 2.44,$$

and so the line is $y = 2.44x$. To estimate y for $x = 7$, we get $2.44(7) = 17.08$ instead of the 21 observed.

Exercises (continued)

4. For the data in Table 2

   a) Graph the data as in Fig. 1.

   b) Draw a line fitting the data points as closely as you can.

c) Use the formula to compute the slope and compare it with the slope of the line you drew in (b).

d) Use the equation of the line fitted in (c) to estimate the total points scored by Hook from his fouls.

e) Now use the equation to estimate the number of Hook's fouls from his total points.

f) Compare the slopes computed from the formula for Tables 1 and 2.

5. Gather the data on fouls and points scored for your own school basketball team and do (a), (b), and (c) as in Exercise 4. Compare the slope for your school with those for Tables 1 and 2.

6. Look in a baseball record book for a particular team in a particular year and investigate, as above, the possible relationship in one of the following:

   a) The number of home runs and the number of strike outs for a batter.

   b) The number of home runs and the number of bases on balls for a batter.

   c) The number of hits and the number of bases on balls for a batter.

Reference

[1] Further discussions of fitting lines to points appear in SBE, Detecting Patterns, Sets 11 and 12.

Examples of Graphical Methods

YVONNE M. M. BISHOP

INTRODUCTION

These four examples illustrate several different ways of organizing data to get at its information. In earlier examples you learned a bit about frequency distributions and averages. In these examples you will learn to use histograms, which we will define for you in the first example of this set, and also scatter diagrams and line graphs. It is often surprising how much information will stand out clearly from a well-drawn picture of the data. The same data may benefit from more than one drawing.

EXAMPLE 1. DIABETIC MICE

Sometimes animals develop the same diseases as humans, and by studying animals we can learn more about a disease. One scientist, for instance, has developed a strain of diabetic mice by crossing two pure strains. (A pure strain is one which has been inbred for many generations so that all the mice in that strain have the same genes.) The scientist wished to know how the diabetic hybrid mice differed from the two normal parental strains, and also whether the two normal strains differed from each other.

It was observed that the diabetic mice were larger and heavier than their parents, and the scientist wished to know whether this was due to an overall

The author is at Children's Cancer Research Foundation, Boston, and Harvard University, Cambridge, Massachusetts.

increase or a disproportionate increase in some organs. Table 1 gives the weights of the mice and of different parts of the body at age 10 months.

Table 1. Weights of 10-mo.-old male mice from two normal strains and the diabetic offspring of crossbreeding*

	Body weight (gm)	Heart (mg)	Liver (mg)	Kidney (mg)
Normal parental strain A	34	210	2240	810
	43	223	2460	480
	35	205	1880	680
	33	225	1970	920
	34	188	1940	650
	26	149	1400	650
	30	172	1470	650
	31	201	2060	560
	31	164	1760	620
	27	188	1690	740
	28	163	1500	600
Normal parental strain B	27	118	1640	640
	30	136	1690	690
	37	156	1980	780
	38	150	1810	660
	32	140	1750	750
	36	155	1770	780
	32	157	1780	670
	32	114	1670	670
	38	144	1980	700
	42	159	2260	720
	36	149	2070	800
	44	170	2530	830
	33	131	1750	640
	38	160	2200	800
Diabetic offspring (F_1 hybrids between strain A and strain B)	42	510†	2300	1030
	44	233	2550	1240
	38	211	2070	1150
	52	264	3450	1280
	48	236	2740	1240
	46	232	2750	1100
	34	210	2080	1040
	44	211	2680	1080
	38	186	2100	870

* Original data from Dr. E. Jones, Children's Cancer Research Foundation, Boston, Massachusetts. Reprinted by permission.

† Assumed to be a measurement error and omitted from graphs.

Information given in such a table is easier to grasp and understand if we express it in certain graphical forms. First we look at the table of liver weights and see that they range between 1400 and 3450 mg, so we decide that if we group them into 200-mg groups we will have about 10 groups, which is usually a convenient number for graphing. The first group will then consist of all mice with liver weights between 1400 and 1599, the next of those between 1600 and 1799, and so on. We see that for strain A there are three in the first group, those with liver weights of 1400, 1470, and 1500. So in Fig. 1(a) we draw a horizontal line segment three units above the axis, running from 1400 to 1600. Similarly, we group the two weights of 1690 and 1760 and draw the next line between 1600 and 1800 two units above the axis, and continue until we have included all the strain A mice. The resulting graph is called a histogram. It gives an idea of how the weights cluster and how much they spread. The area under each segment of the graph is proportional to the number of livers falling into the size-group.

If we plot histograms for the different parts of the body (Figs. 1(a), 1(b), and 1(c)) we see that the diabetic mice seem to have organs that are larger than those of their parents. This, however, does not tell us very much unless we know how the organs of the diabetic mice related to their overall size. So now we try a different kind of graph. For each mouse of all three strains, we plot a point whose horizontal coordinate or abscissa is the body weight and whose vertical coordinate or ordinate is the liver weight. In order to distinguish between the different strains, we use different symbols for each of the three samples. Such a graph (see Fig. 2) is called a scatter diagram. In this case we have a scatter diagram of liver weight against body weight. By looking at our scatter diagram, we find that body weight accounts for the differences in liver size; in Fig. 2 we can draw one straight line that falls through the middle of each sample.

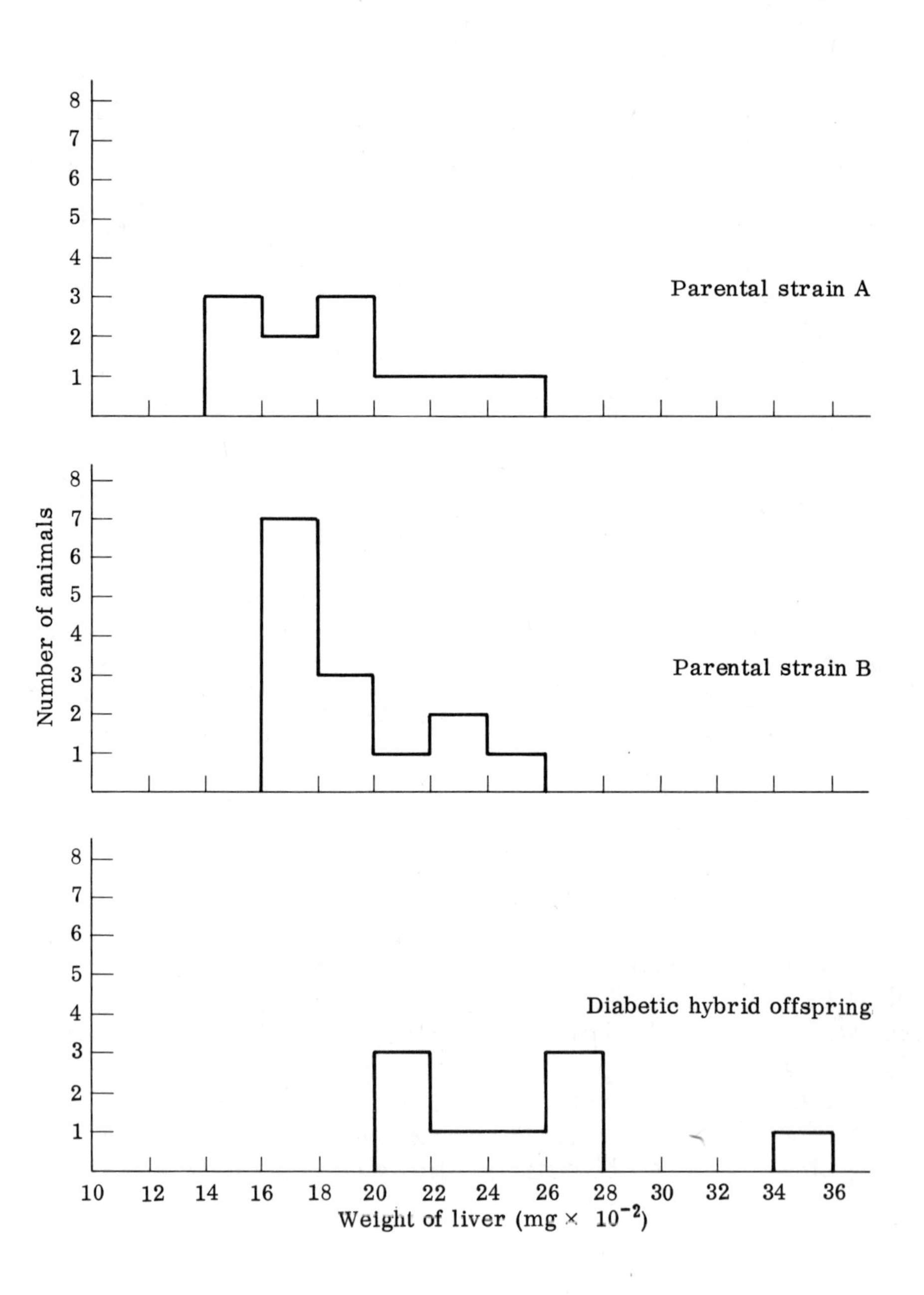
Parental strain A
Parental strain B
Diabetic hybrid offspring
Number of animals
1
2
3
4
5
6
7
8
10
12
14
16
18
20
22
24
26
28
30
32
34
36
Weight of liver (mg × 10^{-2})

Fig. 1(a). Liver

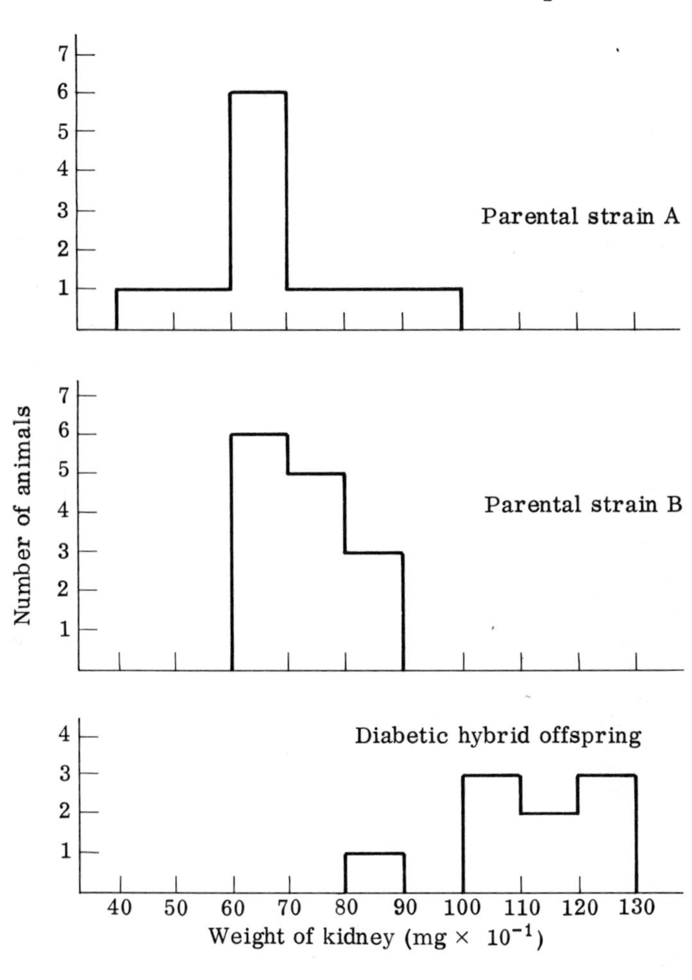

Fig. 1(b). Kidney

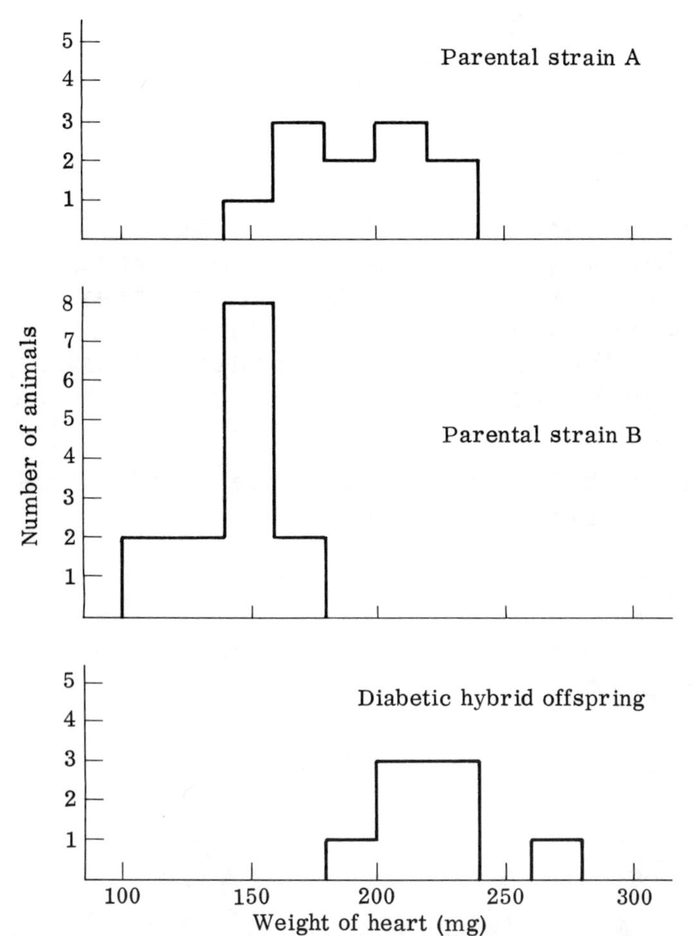

Fig. 1(c). Heart

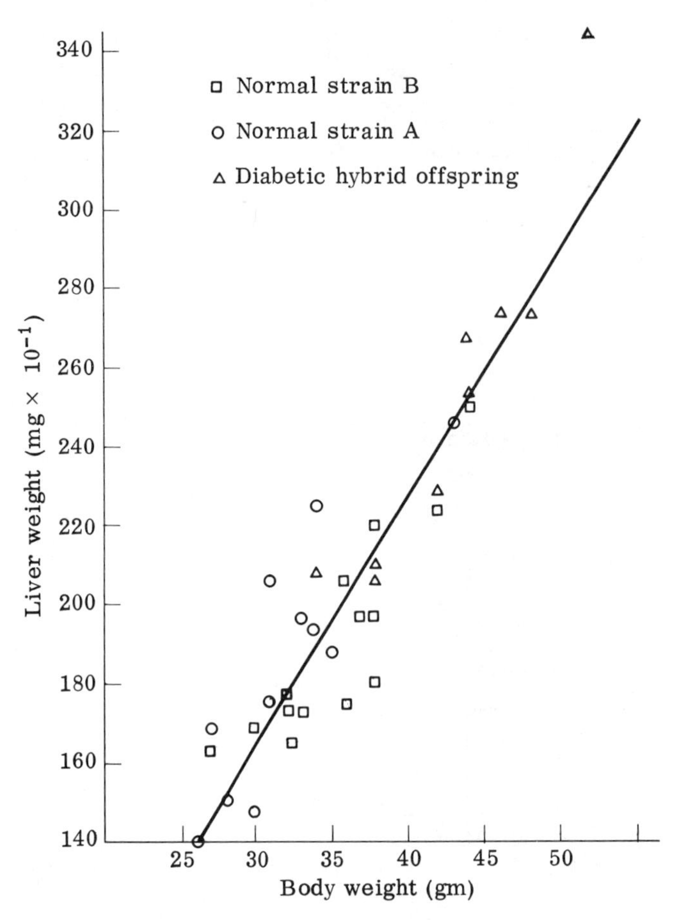

Fig. 2. Liver weight versus body weight.

What about kidneys and hearts? Figure 3 is a scatter diagram for kidney weights against body weights for the two parental strains and Fig. 4 is a scatter diagram for heart weights against body weights. In Fig. 3 we find that for the kidneys the points are more scattered, but the parents again fall on about the same line.

In Fig. 4 we find that for the heart the two parental strains fall on two different lines.

Exercise for Example 1

1. Add the data for the offspring to the graphs in Figs. 3 and 4. What do you find?

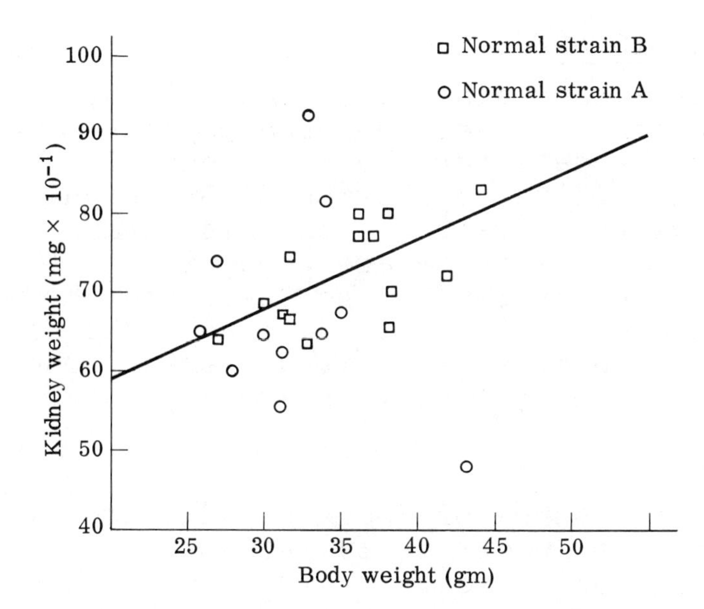

Fig. 3. Kidney weight versus body weight.

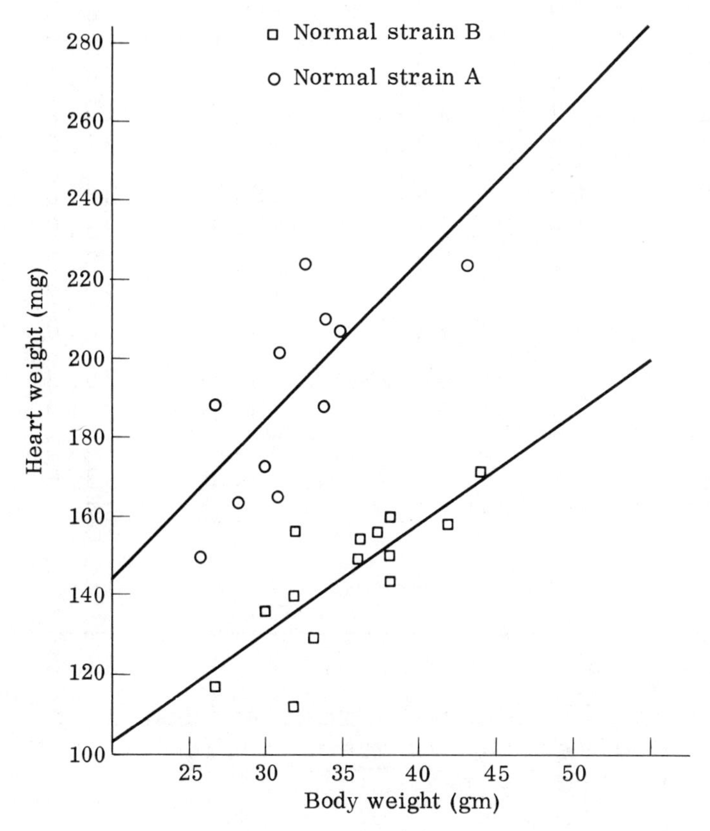

Fig. 4. Heart weight versus body weight.

EXAMPLE 2. HODGKIN'S DISEASE

Doctors studying disease can gain useful insights by studying the distribution of cases in large populations. Table 2 gives the data for the incidence of Hodgkin's disease in the white population of Brooklyn, New York, during the years 1943-1952. The histogram for the total number of patients by age group is plotted in Fig. 5. We see a peak at age 25-30 and another lower peak at age 55-65. This does not help us very much unless we know whether this is different from the age distribution of people living in Brooklyn at that time.

Table 2. Mean annual incidence of Hodgkin's disease in the white population of Brooklyn, 1943-1952, according to sex and age of patients

Age at diagnosis	Number of patients M	F	Total	Incidence per million M	F	Total
0-4	0	0	0	0.0	0.0	0.0
5-9	2	1	3	2.2	1.1	1.7
10-14	3	5	8	3.5	6.2	4.8
15-19	16	18	34	18.3	20.0	19.2
20-24	29	16	45	29.4	15.1	22.0
25-29	27	33	60	26.1	29.3	27.8
30-34	24	24	48	24.3	21.9	23.0
35-39	23	19	42	22.9	17.1	19.9
40-44	26	19	45	26.8	18.7	22.7
45-49	20	12	32	22.7	13.4	18.0
50-54	25	17	42	30.2	20.5	25.3
55-59	37	10	47	54.1	15.0	34.8
60-64	30	17	47	54.7	31.2	43.0
65-69	28	16	44	69.7	36.8	52.6
70-74	20	10	30	82.3	36.1	57.7
75 and over	11	8	19	53.6	30.1	40.3
All ages	321	225	546	25.7	17.4	21.5

Source: MacMahon, Brian, "Epidemiological evidence on the nature of Hodgkin's disease," Cancer, 10.2, No. 5 (1957): 1045-54. Reprinted by permission.

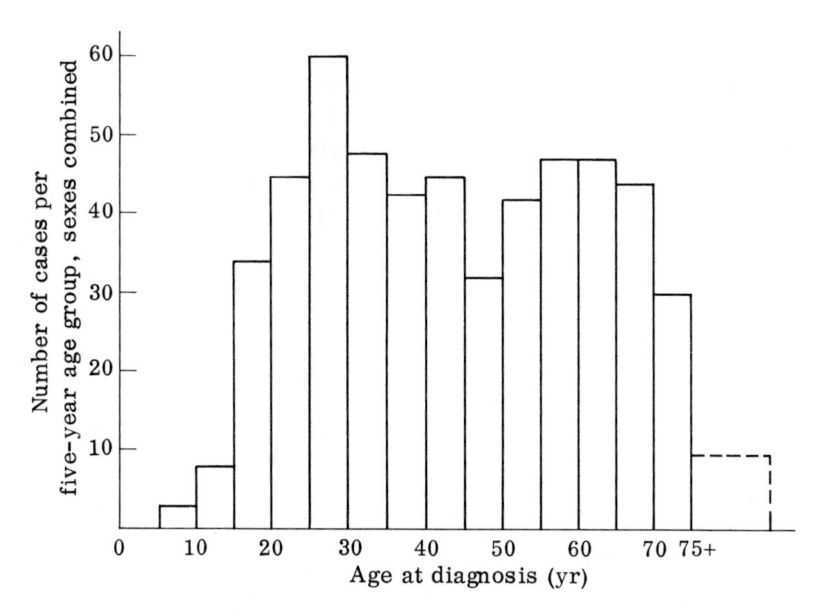

Fig. 5. Total number of Hodgkin's patients in the white population of Brooklyn, 1943-1952.

The authors obtained the population distribution from the census and were then able to compute the incidence rate for each age group. These values are also given in Table 2. To look at rates, we need a different kind of graph known as a line graph. A line graph for the incidence rate for both sexes combined is shown in Fig. 6. It was obtained by plotting a point for each rate at the appropriate height above the abscissa and then joining the points. Thus for the age-group over 5 and under 10 we have a rate of 1.7 cases per million exposed, and as this rate reflects the average risk for this age-group, we plot the point in the middle of the group at age 7 1/2. Then for the age group 10-15 we plot the rate at age 12 1/2. We know that the risk of getting the disease does not change abruptly when a person reaches his tenth birthday, so we join the two points by a line to indicate that between these ages the rate is increasing.

We can now look at the shape of the graph, and again we see two peaks, one between age 25 and 30, as in the histogram, the other in the oldest age group. This double peak is unusual and suggests that there may be two different causes for this disease.

Before we place much confidence in this finding, we must investigate whether it is caused by some peculiarity of the Brooklyn population. If, for instance, the population had a different male/female ratio at different ages, and the disease affected one sex more than the other, we might see such a double peak when we plotted the total population rates, but this would disappear when we looked at the sexes separately. There are other variables that could cause the peaks in the Brooklyn data, so we should also see if the double peak occurs in other populations. If it does occur in other populations, we would also be interested in knowing if it occurs for other diseases.

Fig. 6. Incidence rate of Hodgkin's disease for the white population of Brooklyn, 1943-1952.

<u>Exercises for Example 2</u>

1. Plot the incidence rates for males and females separately. What do you see?

2. Here are the data for Hodgkin's disease in Connecticut (Table 3) and Reticulum Cell Sarcoma in Connecticut (Table 4). Compare the overall incidence rates of these two diseases. Does the Connecticut data for Hodgkin's disease show the same things as the Brooklyn data?

Table 3. Mean annual incidence of Hodgkin's disease in the population of Connecticut, 1950-1959, according to sex and age of patients

Age at diagnosis	Number of patients M	F	Total	Incidence per 100,000 M	F	Total
0-4	2	0	2	0.2	0.0	0.1
5-9	4	2	6	0.4	0.2	0.3
10-14	3	9	12	0.4	0.9	0.7
15-19	18	18	36	2.5	2.5	2.5
20-24	35	12	47	5.2	1.7	3.5
25-29	41	30	71	5.3	3.6	4.5
30-34	29	28	57	3.4	3.2	3.3
35-39	20	20	40	2.3	2.3	2.3
40-44	28	13	41	3.6	1.6	2.6
45-49	33	20	53	4.7	2.8	3.8
50-54	25	16	41	4.0	2.5	3.3
55-59	28	20	48	5.0	3.4	4.2
60-64	23	20	43	4.9	3.9	4.4
65-69	19	14	33	5.2	3.4	4.3
70-74	13	17	30	4.9	5.3	5.1
75-79	18	14	32	11.8	6.4	9.1
80-84	14	3	17	17.1	2.7	9.9
85+	4	5	9	9.2	7.4	8.3

Source: Cancer in Connecticut, Incidence and Rates, 1966. Connecticut State Department of Health.

Table 4. Mean annual incidence of reticulum cell sarcoma in the population of Connecticut, 1950-1959, according to sex and age of patients

Age at diagnosis	Number of patients M	F	Total	Incidence per 100,000 M	F	Total
0-4	1	1	2	0.1	0.1	0.1
5-9	1	0	1	0.1	0.0	0.1
10-14	0	1	1	0.0	0.1	0.1
15-19	1	2	3	0.2	0.3	0.3
20-24	3	0	3	0.5	0.0	0.3
25-29	2	4	6	0.3	0.5	0.4
30-34	8	3	11	0.9	0.3	0.6
35-39	4	3	7	0.5	0.3	0.4
40-44	11	8	19	1.4	1.0	1.2
45-49	11	7	18	1.5	1.0	1.3
50-54	9	6	15	1.4	1.0	1.2
55-59	12	12	24	2.2	2.2	2.2
60-64	17	17	34	3.6	3.4	3.5
65-69	16	19	35	4.2	4.5	4.4
70-74	10	6	16	3.8	1.9	2.9
75-79	14	8	22	9.1	3.9	6.5
80-84	2	4	6	2.5	3.9	3.2
85+	2	6	8	5.5	8.6	7.1

Source: Cancer in Connecticut, Incidence and Rates, 1966. Connecticut State Department of Health.

EXAMPLE 3. POPULATION OF THE UNITED STATES

If we wish to compare histograms of two samples of different sizes, we can plot percentages instead of the counts in each group, so that the total area under the graphs will be the same. In Table 5 we have the age distribution of the population of the United States in 1960 according to where persons were born.

Table 5. Percentage age distribution and the age-specific death rates for the foreign-born and native-born white populations of the United States in 1960

	Population per 100		Death rates per 1000	
Age	Native-born	Foreign-born	Native-born	Foreign-born
Less than 1	2.35	0.07	23.4	17.2
1-4	9.21	0.91	0.9	2.2
5-14	20.31	4.03	0.4	0.6
15-24	13.83	5.04	1.0	1.0
25-34	12.97	8.31	1.2	1.1
35-44	13.77	10.41	2.6	2.1
45-64	19.66	37.72	10.6	12.8
65-74	5.30	22.06	36.1	43.7
75-84	2.19	9.71	87.2	102.3
85 and over	0.43	1.73	210.6	244.3
Total	100.02	99.99	8.2	29.0

Source: U. S. Bureau of the Census.

If we inspect these data, we see that the age-intervals are unequal. The first category covers individuals under 1 year of age, the next category covers those between 1 and 4, and other categories are also for varying numbers of years. This is a common practice because, as we see from the death rates, the risk of dying in the first year of life is very different from the risk at older ages. Before we can plot a histogram, we must find the percentage of the population for each year of life. For instance, for native-born, 9.21% of the population is between 1 and 4. We must plot the line at 9.21/4 = 2.3% and extend it over the four years. If we do not do this and plot the line at 9.2%, the areas under the different segments of the histogram would no longer be proportional to the numbers of people. In this example it would look as if there were four times as many children between 1 and 2

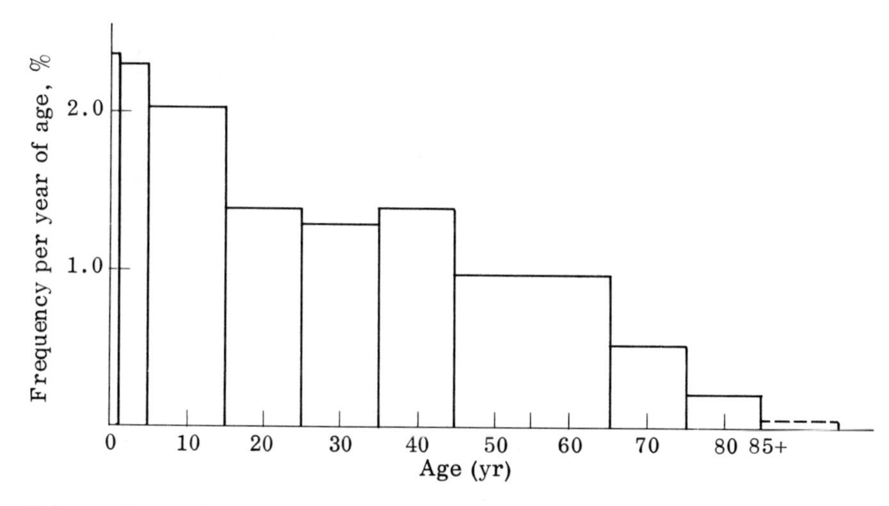

Fig. 7. Percentage age distribution of native-born population of the United States, 1960.

as there were under 1 year, which would be a very strange finding! The histogram for the native-born is given in Fig. 7.

If we look at the death rates in Table 5, we see the crude death rate (in the last line) is much higher for the foreign-born than for the native-born. This rate is computed by taking all the deaths and dividing by the total population. It seems puzzling that the crude rates should be so different, so we look to see if this difference occurs for each age group. The line graph for the native-born is given in Fig. 8; notice that the rates are plotted at the middle of each age group without further adjustment, regardless of the size of the age group. (The age-specific death rate is just the death rate for those in the age group concerned.)

Exercises for Example 3

1. Plot the histogram for the foreign-born population. How do they differ from the native-born? What do you know about immigration laws that could cause this result?

2. Plot the line graph for the foreign-born age-specific death rate. Why is the crude death rate for the foreign-born higher than that for the native-born?

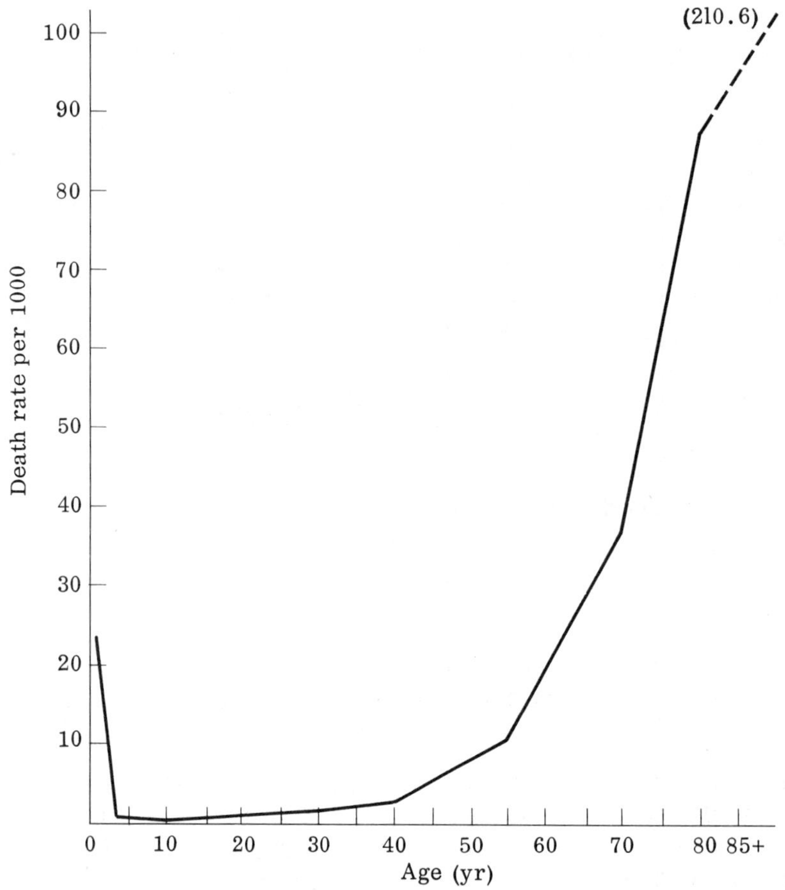

Fig. 8. Age specific death rates for native born population of United States, 1960.

EXAMPLE 4. HEART DISEASE

Table 6 refers to 330 patients with pre-existing symptoms of heart disease for variable periods before their heart attacks.

Table 6. Duration between onset of symptoms of disease and first attack of coronary occlusion

Duration	Number	Duration	Number
1 day	4	4 years	22
2-6 days	12	5 years	16
1-4 weeks	20	6 years	12
1-2 months	28	7 years	8
3-6 months	54	8 years	12
7-12 months	45	9 years	4
1 1/2 - 2 years	50	10 years or more	20
3 years	23		
		Total	330

A first glance at these data might suggest that heart attacks occur most frequently between three months and two years after the onset of heart disease symptoms (from J. Am. Med. Assn., 146 (1951): 998).

Exercise for Example 4

1. Plot the histogram for the first year. Remember to take into account the unequal intervals by computing the number of cases per month first. When did the highest rate occur during this year? What was the total number of cases during the first year? How does this compare with subsequent years? Do you agree with the suggestion that heart attacks occur most frequently between three months and two years after the symptoms?

Babies and Averages

WILLIAM H. KRUSKAL

INTRODUCTION

In Set 3, Example 9, we introduced the idea of average value. The purpose of this example is to look at averages much more carefully than before and to relate them to frequency distributions as well as to the notion of a stable population. Averages used carefully give lots of information; averages used carelessly may give lots of misinformation.

Nearly everyone likes babies, yet most of us know that, sooner or later, people will have to have fewer babies than they now do. Otherwise there will be just too many people for the planet. Opinions differ about how soon the crisis may come, but there is general agreement that birth rates will eventually have to decline.

How far will birth rates have to decline to keep the population from growing indefinitely? At first, it would seem that 2 children for each couple would lead to an exact leveling off. Some couples will have no children, however; others will have more than two. So we see that 2 children per couple on the average might be what is required. Even that is not quite right, because some potential mothers never marry and some children die at early ages. Scientists who study this kind of problem (they are called demographers) agree that an average slightly greater than 2 children per

The author is at the University of Chicago, Chicago, Illinois.

couple would in the long run lead to a stable population size.

One demographer, Ansley J. Coale, discusses 2.25 as a reasonable average number of children per family (in the United States) to reach that goal. He writes in part as follows:

> "An average of 2.25 children does not require that 75% of couples have two children and 25% three, although that would produce the desired average. Another possibility is a nearly even division of family size among zero, one-, two-, three-, four-, and five-child families." [1]

These words correctly suggest that only the average number of children per mother counts in determining population growth; aspects of the distribution of children per mother beyond the average simply do not affect population growth (although such aspects may have social importance in other ways).

The concept of average, therefore, is a central one in the study of population, as indeed it is in nearly all numerical studies, from baseball to price levels, elections to conservation of natural resources.

In the case of population growth, seemingly small changes in the average number of children per mother can have large effects over time. The reason is the same as that behind compound interest--in this case, a slightly larger birth rate becomes magnified as each generation gives rise to the next. Another demographer Conrad Taeuber, puts it this way:

> "Women now say that they plan to have two or three children, but the difference between two and three could be momentous. If women average only two children, we would cease to grow, except as immigration would make up the deficit. If women average three children ... then we would grow at a rapid rate." [2]

There are many prophets who point out the dangers of an indefinitely growing population: dangers of social stress and violence, pollution of air and water, starvation, and other unpleasant consequences.

In this set we study the idea of average as it arises for the number of children in families.

EXAMPLE 1. ARITHMETIC OF AVERAGES

How do we find the average number of children per family? To start, suppose there are 4 mothers; 3 of them have 2 children, and 1 has 3 children. The total number of children in these 4 families is

$$2 + 2 + 2 + 3 = 9.$$

To find the average, we divide the total by the number of families,

$$\frac{9}{4} = 2\frac{1}{4} = 2.25.$$

Note that we could rearrange the arithmetic in the following way, which will soon be more helpful:

$$\frac{2 + 2 + 2 + 3}{4} = \frac{(3\times2) + (1\times3)}{4} = (\frac{3}{4}\times2) + (\frac{1}{4}\times3)$$

$$= (.75\times2) + (.25\times3) = 2.25.$$

We see that the average is a sum of products. Each product is a number of children times the proportion of families with that number of children. For example, three quarters, or .75, of the families in our example have 2 children.

Suppose there were 100 families, of which 75 had 2 children and 25 had 3. The total number of children is

$$\underbrace{2 + 2 + \ldots + 2}_{75\text{ times}} + \underbrace{3 + 3 + \ldots + 3}_{25\text{ times}}$$

or

$$(75\times2) + (25\times3) = 225.$$

The average number of children is then 225 divided by 100, the number of families,

$$\frac{225}{100} = 2.25,$$

the same result as before. Again it is helpful to rewrite the arithmetic as follows:

$$\frac{(75\times2) + (25\times3)}{100} = (\frac{75}{100}\times2) + (\frac{25}{100}\times3)$$

$$= (.75\times2) + (.25\times3) = 2.25,$$

as before. This illustrates how the average depends only on the proportions of families having the different numbers of children, not on the actual numbers of families. In both cases, the proportions were .75 and .25 for 2 and 3 children--in percentages, 75% and 25%; in fractions, 3/4 and 1/4. The average is 2.25 whether there are 4 families or 100 families.

Exercise for Example 1

1. Go through the same arithmetic with 600 families, of which 450 have 2 children and 150 have 3

children. Show that the proportions are the same as above and that the average is still 2.25.

It is useful to show such proportions in tabular form:

Number of children	0	1	2	3	4	5	...
Proportion	0	0	.75	.25	0	0	... .

(The dots mean zero proportions for 6 or more children.) Such a table shows a <u>distribution</u> of number of children. The individual proportions are obtained by taking the number of families with a given number of children and dividing by the number of families. The proportions must be zero or positive, and they must add up to one. The proportions can also be expressed as fractions (0, 0, 3/4, 1/4, 0, 0, ...) or as percentages (0%, 0%, 75%, 25%, 0%, 0%, ...). If we use percentages the total must be 100%. To repeat, we usually avoid working with actual numbers of families, as only proportions count. (Proportions are often called relative frequencies, but in this set we use the term "proportions".)

What we have done so far is to verify Ansley Coale's statement that if 75% of couples have 2 children and 25% have 3, then the average number of children per couple is 2.25. The distribution of the number of children is the important fact, and we have seen that the average number of children for a distribution is obtained by multiplying each number by its proportion, then adding the products.

Of course there are infinitely many distributions. For example, a second distribution suggested by Mr. Coale's words is the distribution with equal proportions for 0, 1, 2, 3, 4, and 5 children. Since there are six possibilities, and they are to have equal proportions, each must have proportion 1/6. We write down the distribution table (using 1/6 here for simplicity instead of a percentage or decimal form).

Number of children	0	1	2	3	4	5	6	...
Proportion	$\frac{1}{6}$	$\frac{1}{6}$	$\frac{1}{6}$	$\frac{1}{6}$	$\frac{1}{6}$	$\frac{1}{6}$	0	...

What is the average? Following our rule, we compute the products,

$$\frac{1}{6} \times 0 = 0 \qquad \frac{1}{6} \times 3 = \frac{3}{6}$$

$$\frac{1}{6} \times 1 = \frac{1}{6} \qquad \frac{1}{6} \times 4 = \frac{4}{6}$$

$$\frac{1}{6} \times 2 = \frac{2}{6} \qquad \frac{1}{6} \times 5 = \frac{5}{6}$$

and add them to get 15/6 = 2.5, slightly higher than the earlier 2.25.

Here is another way to the same answer. Suppose there are 6 families; one has no children, one has 1 child, one has 2 children, and so on up to 5 children. The total number of children is

$$0 + 1 + 2 + 3 + 4 + 5 = 15.$$

To get the average, divide by 6, the number of families. The result is 15/6 = 2.5, as before. This would hold true for 18 families as well, of which 3 have no children, 3 one child, etc. You would again have 2.5 children per family as the average.

Recall Ansley Coale's remark that a nearly even division among family sizes from 0 through 5 would give an average of 2.25. We have discovered that an exactly even division gives the average 2.5. Is there some slight shift in the equal proportions that would give exactly 2.25? Yes; in fact, there are many, infinitely many. An example is

Number of children	0	1	2	3	4	5	6	...
Proportion	$\frac{13}{60}$	$\frac{1}{6}$	$\frac{1}{6}$	$\frac{1}{6}$	$\frac{1}{6}$	$\frac{7}{60}$	0	...

Exercises for Example 1 (continued)

2. Compute the average number of children for this distribution. Your answer should be 2.25.

3. Find one other distribution giving rise to an average of 2.25.

4. Although only the average number of children per mother counts in determining population growth, other aspects of distributions may have important personal and social effects. For example, the following distributions both have 2.25 as their average (the first is the distribution we started with).

Number of children	0	1	2	3	4	5	6	...
Proportion for distribution 1	0	0	.75	.25	0	0	0	...
Proportion for distribution 2	.55	0	0	0	0	.45	0	...

Discuss briefly the social effects of having all families of size 2 or 3 children as against having either no children or 5. (These are extreme cases for simplicity of discussion. We are not considering situations of divorce, remarriage, children born out of wedlock, etc.)

EXAMPLE 2. WHEN IS AN AVERAGE POSSIBLE?

It is interesting to ask what distributions have a specified average, say 2.25. Suppose, for simplicity, that all families have either 1 or 3 children. Let p be the proportion with 1 child, so that 1-p is the proportion with 3 children. The distribution table looks like this:

Number of children	0	1	2	3	4	...
Proportion	0	p	0	1-p	0	...

so that the average number of children per family is

$$(p\times 1) + ((1-p)\times 3) = p + 3(1-p) = 3 - 2p.$$

We ask the question: For what values of p (if any) between 0 and 1 does

$$3 - 2p = 2.25?$$

Exercises for Example 2

1. Show that the conditions hold for exactly one value of p, $p = 3/8$.
2. Carry out a similar analysis if families can have only 1 or 4 children.
3. What happens if we assume that families can have only 1 or 2 children?
4. What general conclusions does this suggest for two possible numbers of children? Do the conditions make intuitive sense? Try to express them in algebraic terms, letting m and n ($m < n$) be the two possible numbers of children. [Hint: Separate into three cases: when $n \leq 2$, when $m \leq 2 < n$, and when $2 < m$.

If we start with more than two numbers of children having positive proportions, and ask what distributions have a specified average, the algebra becomes harder. You can get the flavor of this mathematically interesting kind of question from the following two exercises.

5. Find a distribution in which exactly three numbers of children have proportions greater than zero, and with an average of 2.25.
6. Using the three numbers of children with proportions greater than zero, as above, find a different set of proportions that also gives rise to an average of 2.25. [Caution: Remember that proportions must lie between 0 and 1.]

EXAMPLE 3. A REAL DISTRIBUTION

The distributions we have looked at so far were designed to show general points. They were artificial and simple in order to keep calculations easy. We now look at a real distribution from the United States Census of Population, 1960.

Number of children	0	1	2	3	4	5	6
Proportion	.141	.172	.262	.182	.105	.056	.031

7	8	9	10	11	12 or more
.019	.012	.007	.005	.003	.005

Proportions of Women in the United States 40-44 years old in 1960, having ever had indicated numbers of children. For more detail see Tables 2 and 3 of [3].

By concentrating on women 40-44 years old, most births have already taken place, yet most women are still alive and thus enter the count.

Exercises for Example 3

1. Check that the proportions add to 1. If they do not add to exactly 1 because of rounding off, a footnote to that effect is expected.

2. Round each of the proportions to two decimals, rather than to the three decimal places shown. Follow the usual rule that if the third place is occupied by a 5, you round up or down so as to have an even number in the second place (for example, .105 would round to .10).

   Now add the rounded two decimal proportions. You will find that the sum is slightly less than 1.

   A sensible device to get around part of that problem would be to pool the last three proportions, calling their sum the proportion of women with 10 or more children. Why would this be sensible? Does it lead to a sum of exactly 1? If it does not, but if the sum is very close to 1, most users of the table would be satisfied.

3. Examine the proportions in order from left to right. Notice that they start at 0 children, reach a maximum (.262) at 2 children, and then decline with more children. Does this pattern seem reasonable to you in terms of families you know about?

Graph the proportions on the vertical scale of graph paper against the number of children on the horizontal scale. Make a heavy dot for each point graphed and draw a free-hand curve through the dots. Does the curve seem smooth? Are any points apparently out of line? (The last proportion cannot be graphed, since it doesn't correspond to a specific number of children.)

The above three exercises show the kind of checking of a table that is always good to do but is not always done. Sometimes statistical tables in newspapers and magazines don't add up properly, or they show other signs of sloppiness. Even tables put out by the best statistical organizations may show difficulties.

Other kinds of questions should also be asked when examining a table. For example, the careful student of the above table might wonder how the women were asked about the number of their children, about the possibility that women may not report children who died soon after birth or who were adopted at an early age by other people, and about other misunderstandings or error. The introductory material in the Census publication from which the above table came discusses such issues.

Now we turn to computing the average number of children from the table. As we take the products 0×.141, 1×.172, 2×.262, and so on, all is fine, but how are we going to handle the last category of 12 or more children? We must use it in getting an average, but how do we treat this group for which there is not a definite number of children?

The Census statisticians themselves use an approximation; they increase 12 by 1 to 13, and take 13 as the number of children in those 5 out of 1000 families of the largest category. Of course those large families do not all contain exactly 13 children, but the assumption that they do makes tabulation and calculation easier, without seriously affecting most calculations.

Exercises for Example 3 (continued)

4. Calculate the average number of children for the above distribution. Use the Census approximation, assuming 13 children for the largest category.

5. How sensitive is the average to the approximation you used? See how the average would change if you assumed 12 children for the largest category, or 14. Or some plausible finer breakdown, say

12	13	14
.002	.002	.001

[Hint: You do not have to go back and redo all the computations.]

6. Try to think of some problem for which it would be important to know the whole distribution in detail, all the way through those large families.
7. Construct a real distribution of numbers of children by asking each member of the class to take a family of relatives and give the number of children in it. Put these all together to obtain a distribution for the class. Compare the distribution with the national one reported by the census. Calculate the average for your class distribution, and compare the average with the national one. Discuss possible reasons for the differences.

Your discussion will probably note that families with no children are unlikely to enter into the class distribution, in part because families with no children cannot provide members of your class, and in part because you may be less likely to think of families with no children. Similarly, families with large numbers of children are more likely to enter the class distribution than families with small numbers. Thus these forces will tend to give you a larger average than the national one. On the other hand, you may choose some families that are not yet complete, and therefore may get a smaller average. Other factors are the changes since 1960, and differences among parts of the country.

EXAMPLE 4. IS THERE AN AVERAGE FAMILY?

The average number of children per family is a good statistic for considering population changes, and the things that go with population changes, such as schools, food, roads, water systems, etc. But in our earlier examples no real family could have exactly the average number of children. Obviously no family can have 2.25 children--it isn't even a good joke to talk of a fractional child. Thus it makes no sense to speak of an average family with respect to number of children.

Even if the average were a whole number, we could not speak of an average family. To clarify this, let us take a new distribution.

Number of children	0	1	2	3	4	5	6	...
Proportion	0	.75	0	0	0	.25	0	...

Exercises for Example 4

1. Show that this distribution leads to an average of 2 exactly.

Although the average number of children in this distribution is 2, no family in fact has 2 children.

To be more realistic, let us consider the following distribution.

Number of children	0	1	2	3	4	5	...
Proportion	.15	.20	.30	.20	.15	0	...

2. Show that this distribution also has the average 2.

Consider the families with 2 children from a population for which the above distribution holds. There will presumably be lots of them, for 30% of the families have exactly 2 children and might at first thought be called average families with respect to number of children.

But there is a difficulty. Suppose, for illustration, that a tax on children--to keep the birthrate down--were under consideration. Suppose further that the tax is a progressive one, calculated to be proportional to the <u>square</u> of the number of children. Then we might well be interested in the average of the squared numbers of children. If a two child family is average in a truly useful sense, then we would expect it to be average in terms of squared number of children; that is to say, we would expect the average squared number of children to be 2^2, or 4.

Let us, however, work out the actual average squared number. Write down the table for exercise 2, but with squared numbers of children.

Squared number of children	$0^2=0$	$1^2=1$	$2^2=4$	$3^2=9$	$4^2=16$	...
Proportion	.15	.20	.30	.20	.15	...

Note that the proportions are the same as before. Now compute the average. It is

$$\begin{array}{rcrcr} .15 & \times & 0 & = & 0.0 \\ .20 & \times & 1 & = & .2 \\ .30 & \times & 4 & = & 1.2 \\ .20 & \times & 9 & = & 1.8 \\ .15 & \times & 16 & = & \underline{2.4} \\ & & & & 5.6 \end{array}$$

The average, 5.6, is a number considerably greater than 4. In short, the square of the average number of children is not at all the same as the average squared number of children. This phenomenon makes the use of the average family concept misleading.

Exercises for Example 4 (continued)

3. Carry out similar computations for the cube of the number of children. Use the distribution of exercise 2. Compare $2^3 = 8$ with the average cubed number of children.

If we leave arithmetic and mathematics, the argument against the average family concept may be explained more dramatically. A family that is average in terms of number of children, income, education of parents, and other traits, will nearly always be unusual in some way. The father will be an enthusiast of parachute jumping; or the mother's great-great-aunt was the first woman to graduate from a famous medical school; or the oldest son is regional table-tennis champion; or sister Mary had an almost incurable rare disease that spontaneously went away. Look hard enough--or not even very hard--and you will find something remarkably nonaverage about any person or family. And if, each time an unusual trait turns up, the searcher rejects the family and freshly requires averageness for that trait, he will quickly run out of families.

If there really could be a family that was average on a very large number of traits, it would, by that very characteristic, be highly unusual. So the average family concept is inherently inconsistent. Let us glory in our individualities!

References

[1] Coale, Ansley,J., "Man and his environment," Science 170 (9 October 1970): 132-136.

[2] Taeuber, Conrad F., speech at Mt. Holyoke College, January 1971. Reported in New York Times, 14 January 1971.

[3] U. S. Department of Commerce, Bureau of the Census, United States Census of Population 1960, Final Report, part PC(2)-3A.

Collegiate Football Scores

FREDERICK MOSTELLER

INTRODUCTION

Many studies raise questions which require the collection of data to answer. But after the data have been collected, the main analysis still remains to be done. How do you get answers to the original questions? And after the answers are in, what new questions are suggested by the data? The four examples in this set will start you off on such a search among some data about football, baseball, and hockey scores. You know enough about the sports to ask sensible questions of interest to you and your friends. The patterns you find may have meanings you can discover.

These examples also give another look at frequency distributions and introduce two new kinds of averages, the mode and the median. The first is the "most frequent number" in a set of numbers, and the second is the "middle number" when the set is arranged in order of size. These quantities give a somewhat different kind of information from the average we have looked at before. The examples also show how tables and graphs can produce regularities and irregularities where none seemed apparent before. In the fourth example, frequencies enable us to talk about the "chances" of a team winning.

In the examples, we analyze the 1158 collegiate football scores reported for the year 1967 in the 1968 World Almanac in order to discover their regularities.

The author is at Harvard University, Cambridge, Massachusetts.

The examples are independent in the sense that each may be tackled without having to work through the others in the set.

SCORING IN FOOTBALL

The scoring in 1967 collegiate football allowed 6 points for a touchdown with a bonus opportunity to try for a 1- or 2-point conversion (1 point for kicking a placement after touchdown, or 2 points for a running play conversion after touchdown, the kick being much more frequently chosen), 3 points for a field goal, and 2 points for a safety, a very rare play. Since touchdowns followed by successful 1-point conversions are frequent, we can expect many scores to come in multiples of 7, or nearly that. Field goals, while frequent, are not as frequent as in professional football, partly because the goal posts are 10 yards farther away in the collegiate game.

EXAMPLE 1. FREQUENCY DISTRIBUTION

Given the football scores in the 1968 World Almanac, how would you construct a frequency distribution of winner's versus loser's scores for the 1158 games?

Solution. Since each game produces two scores, a two-way table offers a convenient summary of all the outcomes. Table 1 shows the number of times each pair of scores occurred, with winner's score along the top and loser's down the side. For example, in 6 games the winner had 3 and the loser 0 points, as shown in the box in the upper left-hand corner of the table. Similarly, reading under the column headed 14, we see that 12 games had 14-0 scores, that none had 14-2 scores, but 4 had 14-3.

Exercises for Example 1

1. In Table 1, what does the 3 in the upper left-hand corner of the table mean?
2. In Table 1, which tie involved the highest total score?
3. In Table 1, what does the 219 at the right of the first row mean?
4. In Table 1, what does the 34 at the bottom of the sixth column mean?
5. Why is there a large blank triangle at the lower left of Table 1? ◺

6. Why is there a large blank triangle at the lower right of Table 1? ◺

7. How many ties are there?

8. What is the most frequent losing score? The most frequent winning score?

9. How many ties are there in the hockey scores of Table 7?

10. What does the 82 in the sixth column of the baseball scores tabulated in Table 8 mean?

Solution to Exercise 7. Ties are shown along the diagonal extending southeast from the upper left-hand corner. The 3 in the upper left-hand corner means that 3 games among these 1,158 ended in 0-0 ties during 1967. In 1967, the tie with the most scoring was 37-37 for Alabama vs. Florida State.

Table 1 gives us a way of assessing the rarity of the Harvard-Yale 29-29 tie in 1968. First, ties themselves are rare these days, only 26 in all or about 2%. Only 3 games had ties with higher scores. Looked at another way, in only 20 games did each team score at least 29.

EXAMPLE 2. POPULAR SCORES

Given Table 1, answer the following questions.

a) What are the most popular scores for a game?

b) What are the most popular scores for a winning team?

c) What are the most popular scores for a losing team?

Statisticians call the most popular or most frequent score the modal score. "Popular" does not mean "well-liked" in this discussion.

Solution. (a) Game scores. The most popular score for a game as shown by Table 1 is 14-7, which occurs 16 times, followed closely by 7-0 and 21-7 (with 14 occurrences each), then 10-7, 14-0, 14-13, 17-0, 21-14, with 12 each. Among the higher scoring games, the most frequent are 28-7, 28-14, and 35-0 with frequencies of 10, 11, and 11, respectively. As we anticipated, multiples of 7 are prominent among these popular scores.

(b) and (c) Team scores. The scores that occur with high frequency usually stand out from those near them because of the lumpiness of the scoring system. We can see this even more clearly in the marginal totals of Table 1. These are the numbers in the rightmost column and in the bottom row that give the row and

Table 1. Frequency distribution of 1967 college football scores

Loser \ Winner	0	2	3	4	6	7	8	9	10	11	12	13	14	15	16	17	18	19	20	21	22	23	24	25	26	27	28	29	30	31	32	33	34	35	36	37
0	3		6		4	14	1	3	4		3	5	12	3	7	12		5	10	11	2	4	4	1	5	4	9	5	3	6		3	7	11	1	4
2					1	1									1					1						1										
3			1		1	5		1	1			3	4	1	1	1	1		3	2		1		1	1	1		1					1	1	1	
4																																		1		
6					1	11	1	1	1			5	6	1	3	5		2	1	5	1	4	6		1	4	8		4	3			5	3	2	2
7						3	1	7	12	1	5	8	16	1	2	9	3	4	10	14	3	7	8	1	2	6	10	3	8	3		2	2	3		2
8									2		2	4			1	1		2		2	1	1	3	1	1	1	1		1		1			1		
9								1	3		1		3						1	1		2	1						2	1				1		
10									1		2	4	10	3	2	2		1	4	6			1		2		2			3		1				1
11												1									1														1	
12												5	7			1	1	1	1	4	1	2	3				2	1	1	1			1	2		1
13												4	12	3	2	7	3	1	2	4	2		3		3	4	3	1	3	2			1			1
14													4	3	4	8	1	4	10	12	1	7	7	1	4	7	11	1	2		3	2	6	6	1	1
15															1	3	1		4		1	1	2		1				1	1	1					1
16															3	3	2		4	4	2		1	1		2	1			1				2		
17																			1	5			4		1	2	2						1	2	1	
18																		2	2	2					1								1	1		
19																			2	3							1	1	1	1			1	1		
20																				5	1	1	4		1	4	1	3		2	2		1			
21																				1	4	3	2		2	3	5	1	1			2	1	4		
22																						1	2	1		1			1	1	1		1			
23																						1	2											2		
24																								1	2	1	3			1	1	1				
25																														2						2
26																										1								1		
27																											2			1						2
28																												1	2	3		1		1		
29																														1		1	1			1
30																														1				1		
31																														1		1				
32																																1				
33																																			1	
34																																			1	
35																																		1		1
36																																				
37																																				1
38																																				
Totals	3	0	7	0	7	34	3	13	24	1	13	39	74	15	27	52	12	22	55	82	20	35	53	8	27	42	61	18	30	35	9	15	30	45	9	20

Compiled from data in The World Almanac and Book of Facts, 1968 Centennial Edition, edited by Luman H. Long, published by Newspaper Enterprise Association, Inc., for the Boston Herald Traveler.

38	39	40	41	42	43	44	45	46	47	48	49	50	51	52	53	54	55	56	58	60	61	62	63	65	67	68	69	70	75	77	81	90	Totals
5		1	5	4	2	2			4	2	1		4	1			2	3	2	1		1	1			1	1			2	1	1	219
			1																														6
1	1							1																									36
																																	1
		2	4	4	2		1		1	1	1	3		3	1			3			1									1			114
2	1		6	1	1		1	2	3	2	1			1		2	1	1	1		2			1			1						186
	2	1	1							1															1								32
				2																													19
	1			2																													48
					1																												4
			1	1						1						1			1									1	1				42
	2			2				1	2	2	1				1																		72
2	1		2	5	1	2	1			2	1	1		3							1												128
2			1		1													1															23
			1	1														1															29
		1		1			1		1																								23
2					1				2																								14
1				1		1																											14
1							1				1	1																					29
			1	2	2			1								1																	36
			1	1		1	1			1																							14
			1										1					1															8
1	1							1																									13
1																																	5
																																	2
			2	1		1	1																										10
		1		1								1																					11
																																	4
																																	2
	1																																3
					1						1																						3
																																	1
			1																														2
																																	2
																																	0
				1																													2
		1																															1
18	10	7	28	30	12	7	7	6	13	12	7	6	5	8	2	4	3	10	4	1	4	1	1	1	1	1	2	1	1	3	1	1	1158

column totals. A special tabulation of these in Table 2 shows the frequencies of those scores. Note that the single most frequent winning score is 21, the most frequent losing score is 0, but the most frequent game score is 14-7, whose components do not match either of these. Of course, 21-0 was frequent, and the difference in count between its 11 and the 16 from 14-7 is not so large that it could not be reversed in data from another season. Two important points have been illustrated: (1) the joint distribution of winning and losing scores (here, game scores) may have information different from, indeed here contrary to, that suggested by the separate distributions for the two variables (here, winning and losing scores); (2) the most popular event may be rare--here the most popular score occurs in only 1% of the games. This second point has an important consequence. It says that the joint distribution of two variables each using many categories may not have many cases per cell, even when the total number of cases seems large.

Exercises for Example 2

1. Use Table 6 to check whether 14-7 persisted as more frequent than 21-0 for the 1968 season.
2. Use Table 6 to find the modal winning and losing scores for the 1968 season. Do they agree with those for 1967?
3. For the hockey scores of Table 7 find out whether the most frequent game score has components which are the modal winning and losing scores.
4. For the baseball scores of Table 8 find out whether the most frequent game score has components which are the modal winning and losing scores.

MEANS AND MEDIANS

In some of our work we shall need the notion of a weighted mean or average. What is the average losing score for a winning scores of 7? In Table 1 we see that the losing score of 0 occurred 14 times, of 2 once, of 3 five times, of 6 eleven times, and of 7 three times. What we must do is get the grand total of these scores and divide by the number of them in order to get the average:

$$\frac{0\times 14 + 2\times 1 + 3\times 5 + 6\times 11 + 7\times 3}{14 + 1 + 5 + 11 + 3} = \frac{104}{34} = 3.1.$$

The median of a set of scores is the number such that at least half the scores are as large as it or larger and at least half as small as it or smaller. If there is an interval with this property, we take the middle number. For example, given the 9 scores 2, 2, 3, 3, 4, 4, 5, 5, 5, since the fifth score counting from the bottom is 4, the median is 4. The median of the two scores 4 and 7 is 5.5.

EXAMPLE 3. TYPICAL SCORES

Find the average winning and losing scores and the median winning and losing scores.

Solution. By a lengthy calculation in the margins of Table 1, we find that the average winning score is 26.7, the average losing score is 10.0, for an average difference of 16.7 points. And so in round numbers the average game score is 27-10, as opposed to the most frequent score which is 14-7. It is easier to find medians from Table 1. The median winning score is 24, the median losing score is 8, and so 24-8 might be called a median, or middling, game. Thus means and medians gave similar answers in this problem, but this does not always happen.

As a matter of general interest, Table 5 gives average winning and losing scores in various collegiate football leagues.

Table 3 gives the average winning score associated with losing scores.

Exercises for Example 3

1. Use Table 1 to compute the average losing score when the winning score is 9. Check your answers in Table 4.
2. Find from Table 8 the median winning and losing baseball scores.
3. Find from Table 8 the average losing baseball score when the winning score is 3.
4. Find from Table 7 the average losing score in the hockey games.
5. Find from Table 7 the average winning score in the hockey games.
6. Use the data found in Exercises 4 and 5 to get the average difference between winner's and loser's scores.

EXAMPLE 4. THE CHANCE THAT A GIVEN SCORE WINS

Given the frequencies in Table 1, answer the following:

a) Given a score, say 10, how often does this score win?

b) Are some scores for their size relatively good or relatively poor?

Solution. We consider each team score and ask: "If this score occurs, how often is it a winning score?". The basic data from the margins of Table 1 have been conveniently reassembled in Table 2. The relative frequency that a 10 is a winning score is $24/72 = 1/3 \approx 0.33$. We plot this point (10,0.33) in Fig. 1 to display the result more clearly.

Table 2. Distributions of team scores up to scores of 29

	Frequency				Frequency		
Score	Winning	Losing	Total	Score	Winning	Losing	Total
0	3	219*	222*	17	52*	23	75
2	0	6	6	18	12	14	26
3	7	36	43	19	22	14	36
4	0	1	1	20	55*	29	84
6	7	114*	121*	21	82*	36	118*
7	34	186*	220*	22	20	14	34
8	3	32	35	23	35	8	43
9	13	19	32	24	53*	13	66
10	24	48*	72	25	8	5	13
11	1	4	5	26	27	2	29
12	13	42	55	27	42	10	52
13	39	72*	111*	28	61*	11	72
14	74*	128*	202*	29	18	4	22
15	15	23	38	⋮	⋮	⋮	⋮
16	27	29	56				
				Total	1158	1158	2316

* Six most frequent scores in the complete column.

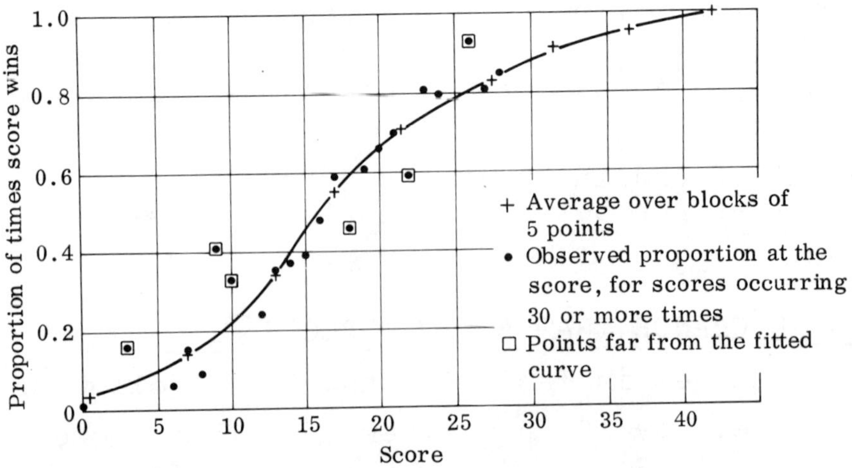

Fig. 1. Proportion of times a given score wins.

The lazy S-shaped curve in Fig. 1 shows the general rise in probability of winning with increasing score. It is a freehand curve that passes through the +'s which are based on average results for 5 scores. In each 5-point interval 0-4, 5-9, 10-14, and so on, we computed the average score and the percentage of wins and plotted the +'s at these coordinates. For example, in the interval 5-9, the average score is (from Table 2)

$$\frac{6\times 7 + 7\times 34 + 8\times 3 + 9\times 13}{7 + 34 + 3 + 13} = \frac{421}{57} = 7.4$$

and the proportion of wins is 57/408 ≈ 0.14. Hence the + point has coordinates (7.4,0.14). Similar calculations are needed for other intervals.

From the smooth curve we see that a score of 16 has a 50-50 chance of being a winner. A team with 40 points is practically assured a victory.

The dots, which correspond to single scores, generally fall close to the freehand curve. The 6 boxed dots farthest from the curve are instructive. We study them to see whether they "make a pattern"--whether we have reasonable explanations for their departures from the curve. There need not be any explanation; after all, some points are bound to be further away than others.

A score of 9, which the freehand curve predicts will win only 18% of the time, actually wins 41% of the time, and thus it is a special score. This produces a large departure between the prediction and the outcome. This departure is called a <u>residual</u>; here it is a difference of 41-18 = 23. The scores 3, 9, 10 are favorable scores. That is, the graph suggests that teams with these scores are considerably more likely to win than the size of the score alone indicates (this is what the freehand curve tells), an additional 9% for 3, 23% for 9, 11% for 10. These are all scores intimately related to field goals. The score 26 seems also to be especially good, 12% above the curve, though perhaps it is due to chance fluctuations.

The unlucky scores are 18 and 22, scoring 14% and 13% below the freehand curve for scores near this size. Of course, Notre Dame once won a famous game with 18 points, but that score generally means missing 3 points after touchdown, which is rather a bad sign.

Exercises for Example 4

1. According to the freehand curve of Fig. 1, what score gives a 20% probability of winning?
2. Use Table 6 to check the finding about winning scores of 3, 9, and 10 for 1968 scores. To do

this, construct the 3 left-most + points corresponding to Fig. 1 for the 1968 data and pass a smooth curve through. Then look at the residuals and discuss.

3. Project. Make a graph corresponding to Fig. 1 for the baseball data of Table 8, but adapted for the baseball data. Discuss similarities and differences from Fig. 1.

Table 3. Average winning score for each losing score

Losing score	Number of games	Average winning score	Losing score	Number of games	Average winning score
0	219	26.7	21	36	30.8
2	6	19.7	22	14	33.6
3	36	19.0	23	8	36.1
4	1	35.0	24	13	31.3
6	114	27.8	25	5	34.8
7	186	24.5	26	2	31.0
8	32	25.1	27	10	37.4
9	19	21.8	28	11	34.7
10	48	20.4	29	4	33.8
11	4	28.5	30	2	33.0
12	42	27.0	31	3	34.3
13	72	24.3	32	3	41.7
14	128	27.8	33	1	36.0
15	23	27.4	34	2	38.5
16	29	24.6	35	2	36.0
17	23	29.2	37	2	39.5
18	14	30.6	38	1	40.0
19	14	29.6			
20	29	28.9	Total	1158	26.7

Table 4. Average losing score for each winning score

Winning score	Number of games	Average losing score	Winning score	Number of games	Average losing score
0	3	0	28	61	11.2
3	7	.4	29	18	10.6
6	7	1.6	30	30	10.6
7	34	3.1	31	35	14.4
8	3	4.3	32	9	16.8
9	13	5.2	33	15	15.9
10	24	6.1	34	30	10.1
11	1	7.0	35	45	11.6
12	13	6.2	36	9	13.8
13	39	7.4	37	20	14.8
14	74	7.9	38	18	11.1
15	15	8.5	39	10	13.1
16	27	7.8	40	7	14.7
17	52	8.3	41	28	10.8
18	12	11.3	42	30	12.8
19	22	8.3	43	12	12.6
20	55	9.6	44	7	13.7
21	82	10.4	45	7	16.1
22	20	12.4	46	6	12.5
23	35	10.5	47	13	8.2
24	53	12.0	48	12	9.7
25	8	11.8	49	7	13.1
26	27	11.3	50	6	13.3
27	42	12.1	⋮	⋮	⋮
			Total	1158	10.0

Table 5. Average for league games

League or conference	Average winner's score	Average loser's score	Average winner's minus loser's	Number of games
Ivy	30.6	11.0	19.6	28
Yankee	21.6	8.9	12.7	15
Middle Atlantic	24.1	10.6	13.6	11
College Division:				
North	27.1	10.2	16.9	21
South	28.1	10.3	17.7	30
Atlantic Coast	22.8	9.5	13.3	25
Big Eight	22.2	7.6	14.7	27
Southeastern	25.7	9.4	16.4	27
Mid-American	24.2	9.0	15.2	21
Southwest	22.7	11.4	11.3	26
Southern	27.0	12.6	14.4	17
Missouri Valley	25.8	10.6	15.2	11
Western Athletic	33.7	15.5	18.1	23
Big Ten	23.9	12.0	11.9	35
Pacific Eight	24.7	8.6	16.0	25

Table 6. Frequency distribution of 1968 college football scores

Losing score \ Winning score	0	2	3	4	5	6	7	8	9	10	11	12	13	14	15	16	17	18	19	20	21	22	23	24	25	26	27	28	29	30	31	32	33	34	35	36	37
0	5		2			4	4		3	5		2	2	8		6	7	2	3	4	7		1	6	2	6	2	12	3	2	8	1	2	5	6	3	3
2							2																														
3						3	4		1	3		1	1	2			2	1	2	2	2			1		2		1	1	1	3		1		1		
4																																					
5																								1													
6							10			4		2	2	6		3	4	1		5	7	1	2	1		2	4	3	1	2	3	4	2	2	6		2
7							5	3	3	6		1	6	14	1	7	4	3		4	12	4	4	7	1	2	6	9	5	4	6	2	2	4	3		5
8										4			3			2			1				1			2		1		1	1		1	2		1	
9										4		1	3	1		1				1		2	1	2	1		1	2			2	2		1	1		
10										2	1	2	6	5			6			6	2			1	1	1	1	2			1		2	1		1	
11														1							1			1		1	1			1	1						
12													1				5			3	5				1		2	8		1	2		1	1	3	2	1
13													2	4	1	2	7	1	3	4	4	3	4	4	1	1	7	6		2	1	1		2	6		3
14														5	1	4	7	2	1	7	8	1	5	7	1	2	7	10		4	8	3	1	3	5	2	3
15																1	3	1		2	1		1		1		2	3				1		1	2		1
16																	2			2	3		2					1		2			1	1	1		1
17																	1	1	2	6	5	1		3		1		2	2				1	1	4		
18																			1				1					2		1	1						
19																					3	2	2	2		1	2	1			2			1	1		
20																				2	9		5	3	2	3	7	2	1	4	2		1	1	4		
21																					1	4	3	4	1	1	1	2		2	1			1	1		
22																								1	1	1	1	2		1	1			1			1
23																										2	1	2					1			1	
24																								1	1	1	3	1	2		1	1	2				
25																											1				1			2	1		
26																											2		2								
27																												2			1	2	1	2	1		
28																												2		3	2			2	2		2
29																													1						1		1
30																																					
31																																	1	1			
32																																		1	2		
33																																			1		
34																																			1		
35																																			1		
36																																					
37																																					
38																																					
40																																					
42																																					
45																																					
47																																					
48																																					
Totals	5	-	2	-	-	7	25	3	7	28	1	9	26	46	3	26	48	12	13	48	70	18	32	45	14	29	51	76	18	31	46	17	20	36	54	10	23

Compiled from data in The World Almanac and Book of Facts, 1969 Edition, edited by Luman H. Long, published by Newspaper Enterprise Association, Inc., for the Boston Herald Traveler.

38	39	40	41	42	43	44	45	46	47	48	49	50	51	52	53	54	55	56	57	58	59	60	61	62	63	64	65	66	68	69	71	76	77	100	Totals
3	2	4	4	2	1	1	1	4	3	3	3		1	2					1	1	1					1	1	1	1	1		1			169
																																			2
1	1			1						1		1													1								1		42
																																			-
																																			1
1	1		2	3	1	3	1	1	1	2	1						1			3									1					1	102
3		2	1	6	4	1	3		1	2		1				1	1	1							1				2						163
1	1				1	1				2										1				1											28
																											1								27
														1																					42
																																			7
1		1		2					1																										41
		1	4	2	1		1		1	1							1	1	1	1															84
3			6	2	1	2	1	3	4	1	3	1																		1					125
1	1					1			1		1																								25
						1			1	1																									19
			1	2								1																							34
	1	1		1						1	1											1													12
			1	2						1		1									1	1													24
1		2		1				2	1		1	1			1			1						1					1						59
		1		2			5	1	3									1			2														37
	1	2	1	1																															15
1				1				1																											10
				1													1																		15
								1											1	1															8
			1	2					1																										8
		1	1						2			1																							14
				2			1				2		1			1									1										21
1	2																						1												7
								1						1																					2
		1								1																									4
																	1																		4
																															1				2
2																	2												1						6
1																				1															3
		1													1																				2
1																																			1
															1					1															2
				1								1																							2
												1																							1
																		1																	1
																									1										1
																						1													1
21	10	17	22	34	9	10	13	14	20	16	12	9	2	4	3	2	7	5	3	9	4	3	1	2	4	1	2	1	6	2	1	1	1	1	1173

Table 7. 1967-1968 Professional hockey scores

Losing score \ Winning score	0	1	2	3	4	5	6	7	8	9	Total
	Frequency distribution of scores										
0	2	14	16	16	10	5	7	2	1	0	73
1		14	26	42	22	17	9	4	2	1	137
2			28	28	38	16	9	6	1	2	128
3				22	14	22	9	1	0	3	71
4					6	5	7	5	0	1	24
5						3	1	3	1	0	8
6							1	0	2	0	3
Total	2	28	70	108	90	68	43	21	7	7	444

Source: Scores provided by Ron Andrews, Publicity Director, National Hockey League.

Table 8. Frequency distribution of baseball scores for the American League in 1968

Losing score \ Winning score	1	2	3	4	5	6	7	8	9	10	11	12	13	14	15	16	Totals
0	38	35	28	20	8	7	8	2	3	2		2	1				154
1		66	42	42	28	17	7	9	3	3	1	2	2				222
2			61	41	21	14	20	8	3	7	3	2				1	181
3				56	24	20	13	7	3	2	3	1		1			130
4					29	11	6	10	1	2	1		1				61
5						13	6	4	3	1		1					28
6							13	2	2		3	1					21
7								2	2								4
8									1	2		1	1				5
9										2							2
10											1		1				2
Totals	38	101	131	159	110	82	73	44	21	21	12	10	6	1		1	810

Source: Scores from Official Baseball Guide for 1969. Chris Roewe and Paul Macfarlane, editors; C. C. Johnson Spink, publisher. Published by The Sporting News, St. Louis.

Reference

[1] These football scores are discussed further in SBE, Weighing Chances, Set 11.

Ratings of Typewriters

FREDERICK MOSTELLER

INTRODUCTION

Here is a large set of numbers; the data, in fact, consist of 625 numbers. These numbers depend on two things, the model of the typewriter being used and the particular typist using it, and so a rectangular array is a good way to display the data. This set will show you how to get more information out of the data by a systematic rearrangement and by studying the averages of the rows and columns of the array.

EXAMPLE 1. RATINGS OF TYPEWRITERS

Twenty-five typists were trained to rate typewriters--they assign a typewriter a rating of 1, 2, 3, 4, or 5, 5 being very satisfactory, and 1 being very unsatisfactory. Each typist rated the same 25 typewriters, one each of 25 different brands. They used them to type several different kinds of things--letters, envelopes, postcards, tables, and so on. When they were finished, their ratings produced the data shown in Table 1. Rank the typewriters from least liked to best liked, and rank the raters from tough raters to easy raters.

Solution. One way to rank the brands of typewriters and raters is to use the sums of their scores, shown in

---

The author is at Harvard University, Cambridge, Massachusetts. Data was kindly provided by Consumers Union, publisher of Consumer Reports, and was brought to the author's attention by Cuthbert Daniel.

Table 1. Typewriter ratings from a low of 1 to a high of 5 given by 25 trained raters to 25 brands of typewriters

Brand	Rater 1	2	3	4	5	6	7	8	9	10	11	12	13	14	15	16	17	18	19	20	21	22	23	24	25	Total
1	3	3	3	3	3	4	3	2	4	4	3	2	4	3	5	3	5	4	5	4	3	5	2	3	4	87
2	4	5	3	3	4	5	4	4	3	4	5	3	3	5	5	5	3	4	4	4	4	5	4	3	4	100
3	4	3	2	3	3	5	2	3	4	3	2	2	4	2	3	4	5	4	3	4	3	3	3	4	3	81
4	2	2	1	1	2	1	1	2	3	2	1	1	1	2	1	3	2	1	2	2	1	1	1	2	1	39
5	4	4	3	4	3	5	3	4	4	3	4	3	2	5	5	4	5	4	2	5	4	4	3	4	3	94
6	2	4	2	2	2	3	2	3	3	2	3	3	2	3	3	3	2	2	3	2	2	4	2	2	2	63
7	1	1	1	1	1	1	1	1	1	1	1	1	2	1	1	1	1	1	1	1	1	1	1	1	1	26
8	2	2	1	2	3	3	1	1	4	3	3	2	2	4	3	4	2	3	4	2	3	2	2	1	3	62
9	2	4	4	3	3	2	4	4	4	4	4	2	1	5	5	3	3	5	5	4	3	3	4	1	4	86
10	3	5	4	3	3	4	5	4	3	4	5	2	4	4	5	3	4	2	5	5	4	5	4	3	3	96
11	1	4	3	4	1	4	3	3	5	5	5	3	5	1	5	4	4	4	2	4	4	4	3	3	2	86
12	2	4	3	2	3	5	4	4	4	3	3	3	2	2	3	2	4	2	3	3	3	3	3	2	3	75
13	2	2	1	1	2	1	1	1	2	2	1	1	1	1	2	2	3	1	1	1	2	2	2	2	1	38
14	2	2	2	4	3	3	2	3	4	3	3	1	2	4	2	4	3	1	2	4	2	3	2	3	2	66
15	3	4	4	4	4	5	5	4	5	4	5	4	3	5	5	4	4	4	5	4	5	5	4	4	5	108
16	3	4	4	3	3	3	2	1	4	2	2	2	2	4	4	3	2	1	2	4	4	2	4	3	4	72
17	4	3	4	4	3	5	4	4	5	4	3	3	5	5	4	4	5	5	5	5	3	4	4	3	3	101
18	1	2	2	2	2	4	1	3	1	3	2	2	2	3	2	3	1	1	3	2	1	1	2	1	2	49
19	3	3	3	5	3	4	3	4	4	3	4	2	3	4	4	3	5	4	3	4	3	4	3	3	4	88
20	3	2	3	2	3	5	4	4	4	4	4	4	3	2	4	4	3	4	4	4	5	4	3	2	4	88
21	4	4	3	4	3	5	2	2	2	2	3	1	3	4	1	2	2	3	2	3	1	3	3	3	2	67
22	2	4	2	4	2	5	4	2	5	3	3	2	3	3	3	4	2	2	5	4	2	3	2	2	2	75
23	4	4	4	5	3	5	5	4	4	5	5	3	4	5	5	5	3	4	5	3	5	3	5	3	5	106
24	3	3	2	5	3	4	4	4	3	3	5	3	4	4	4	5	4	4	4	5	3	3	3	3	4	92
25	4	2	4	2	3	3	3	5	2	4	3	2	2	3	3	4	4	4	4	5	2	5	4	3	4	84
Total	68	80	68	76	68	94	73	76	87	80	82	57	69	84	87	86	81	74	84	88	73	82	73	64	75	1929

the right-hand column and bottom row of Table 1. Another way might be to find the average score given by each rater and the average score received by each brand of typewriter.

Exercises for Example 1

1. Compute the average scores for the highest rated and lowest rated brands.
2. Compute the average score given by the easiest and hardest rater.
3. Compute the grand average rating.

While such computations for every row and column are a lot of work in a table of this size, they are not so much trouble in small tables. Also, these calculations are something a high speed computer can do readily. Brands 10, 2, 17, 23, 15 (see Table 2) are very well liked, but brands 7, 13, 4 get extremely low ratings. Raters 12, 24, 5, 3, 1, 13 give generally low ratings, while raters 16, 15, 9, 20, and 6 tend to give high ratings. The average rating is 3.1.

EXAMPLE 2. ORDERLINESS OF RATINGS

To help you see the information in Table 1 we might try some rearrangement of the tables. One possibility is to rearrange the entire table so that rows are arranged according to totals, and similarly for columns. When this is done we get Table 2. Describe how the main body of Table 2 appears to the eye as compared with Table 1.

Discussion. Although Table 1 looks like a jumble of numbers, with only a little order, Table 2 appears much more orderly, with the upper left having generally small numbers and the lower right having large numbers. Much more information can be got from these tables, as you will see if you read the more complete discussion of this example in reference [1].

Reference

[1] SBE, Weighing Chances, Set 10.

Table 2. Typewriter ratings of Table 1 rearranged in order of row and column totals to show the basic orderliness of the ratings

	Rater																										
Brand	12	24	5	3	1	13	23	21	7	18	25	8	4	10	2	17	22	11	19	14	16	15	9	20	6	Total	Avg.
7	1	1	1	1	1	2	1	1	1	1	1	1	1	1	1	1	1	1	1	1	1	1	1	1	1	26	1.0
13	1	2	2	1	2	1	2	2	1	1	1	1	1	2	2	3	2	1	1	1	2	2	2	1	1	38	1.5
4	1	2	2	1	2	1	1	1	1	1	1	2	1	2	2	2	1	1	2	2	3	1	3	2	1	39	1.6
18	2	1	2	2	1	2	2	1	1	1	2	3	2	3	2	1	1	2	3	3	3	2	1	2	4	49	2.0
8	2	1	3	1	2	2	2	3	1	3	3	1	2	3	2	2	2	3	4	4	4	3	4	2	3	62	2.5
6	3	2	2	2	2	2	2	2	2	2	2	3	2	2	4	2	4	3	3	3	3	3	3	2	3	63	2.5
14	1	3	3	2	2	2	2	2	2	1	2	3	4	3	2	3	3	3	2	4	4	2	4	4	3	66	2.6
21	1	3	3	3	4	3	3	1	2	3	2	2	4	2	4	2	3	3	2	4	2	1	2	3	5	67	2.7
16	2	3	3	4	3	2	4	4	2	1	4	1	3	2	4	2	2	2	2	4	3	4	4	4	3	72	2.9
22	2	2	2	2	2	3	2	2	4	2	2	2	4	3	4	2	3	3	5	3	4	3	5	4	5	75	3.0
12	3	2	3	3	2	2	3	3	4	2	3	4	2	3	4	4	3	3	3	2	2	3	4	3	5	75	3.0
3	2	4	3	2	4	4	3	3	2	4	3	3	3	3	3	5	3	2	3	2	4	3	4	4	5	81	3.2
25	2	3	3	4	4	2	4	2	3	4	4	5	2	4	2	4	5	3	4	3	4	3	2	5	3	84	3.4
11	3	3	1	3	1	5	3	4	3	4	2	3	4	5	4	4	4	5	2	1	4	5	5	4	4	86	3.4
9	2	1	3	4	2	1	4	3	4	5	4	4	3	4	4	3	3	4	5	5	3	5	4	4	2	86	3.4
1	2	3	3	3	3	4	2	3	3	4	4	2	3	4	3	5	5	3	5	3	3	5	4	4	4	87	3.5
20	4	2	3	3	3	3	3	5	4	4	4	4	2	4	2	3	4	4	4	2	4	4	4	4	5	88	3.5
19	2	3	3	3	3	3	3	3	3	4	4	4	5	3	3	5	4	4	3	4	3	4	4	4	4	88	3.5
24	3	3	3	2	3	4	3	3	4	4	4	4	5	3	3	4	3	5	4	4	5	4	3	5	4	92	3.7
5	3	4	3	3	4	2	3	4	3	4	3	4	4	3	4	5	4	4	2	5	4	5	4	5	5	94	3.8
10	2	3	3	4	3	4	4	4	5	2	3	4	3	4	5	4	5	5	5	4	3	5	3	5	4	96	3.8
2	3	3	4	3	4	3	4	4	4	4	4	4	3	4	5	3	5	5	4	5	5	5	3	4	5	100	4.0
17	3	3	3	4	4	5	4	3	4	5	3	4	4	4	3	5	4	3	5	5	4	4	5	5	5	101	4.0
23	3	3	3	4	4	4	5	5	5	4	5	4	5	5	4	3	3	5	5	5	5	5	4	3	5	106	4.2
15	4	4	4	4	3	3	4	5	5	4	5	4	4	4	4	4	5	5	5	5	4	5	5	4	5	108	4.3
Total	57	64	68	68	68	69	73	73	73	74	75	76	76	80	80	81	82	82	84	84	86	87	87	88	94	1929	
Avg.	2.3	2.6	2.7	2.7	2.7	2.8	2.9	2.9	2.9	3.0	3.0	3.0	3.0	3.2	3.2	3.2	3.3	3.3	3.4	3.4	3.4	3.5	3.5	3.5	3.8		3.1
Dev.	-.8	-.5	-.4	-.4	-.4	-.3	-.2	-.2	-.2	-.1	-.1	-.1	-.1	.1	.1	.1	.2	.2	.3	.3	.3	.4	.4	.4	.7		

The Cost of Eating

WILLIAM H. KRUSKAL

INTRODUCTION

Again we have a table of data, this time prices of groceries at different stores. Again we arrange the data in a two-way or rectangular array, since the prices depend on the food item bought and also on the store in which it was bought. Our purpose here is to find out where groceries are most economical. Various problems arise: some stores do not carry certain items, one store may charge more for some things and less for others, prices change on weekends, and so on. Looking at such problems with a statistical eye may help you make better decisions.

Almost everyone talks about the price of food, and most food shoppers compare prices in different grocery stores. Naturally there is curiosity about overall comparisons among stores; we will look at an important way of making such comparisons and at some problems that arise.

A neighborhood newspaper in Chicago, Hyde Park-Kenwood Voices, surveys local grocery store prices from time to time as a service to readers. Here is a small extract from the newspaper's survey report for September 1969. (Table 1 gives the full report.) Numbers in the extract show prices in ordinary dollar and cents units.

The author is at the University of Chicago, Chicago, Illinois.

	Store			
	Mr. G	A&P	Coop	Pete's
Milk, $\frac{1}{2}$ gal. 2%	.58	.59	.57	.67
Eggs, 1 doz. large	.65	.69	.57	.65
Hot dogs, 1 lb.	.79	.95	.85	.79
Tunafish, $6\frac{1}{2}$ oz., canned	.37	.39	.39	.43
Tang®, 18 oz.	.98	.88	.89	.98

As we look at these prices, we see that the Coop has the lowest prices for milk and eggs, but for hot dogs and tunafish the Coop's prices are not the lowest. In the case of Tang, the Coop's price is just one cent more than the lowest. Pete's, to take another store, has high prices for milk, tunafish, and Tang; for eggs, Pete's is intermediate, and for hot dogs, Pete's ties for lowest price with Mr. G.

Looking at the table in another way, we see that the economy-minded shopper, if restricted to these four stores and with plenty of time, would buy milk and eggs at the Coop, hot dogs at Mr. G or Pete's, tunafish at Mr. G, and Tang at the A&P.

So it seems a little complicated to make an overall comparison of the price levels of the four stores. One good way to make such a comparison is to choose a selection of foods, that is, to choose a hypothetical market basket or shopping list, and to find out what it would cost at each store. To illustrate simply, let's take two possible market baskets; we'll call them the Milk and Eggs basket, and the Meat and Fish basket. Here they are.

Milk and Eggs	Meat and Fish
3 half gallons of milk	no milk
2 doz. eggs	no eggs
1 lb. hot dogs	4 lb. hot dogs
no tunafish	2 $6\frac{1}{2}$ oz. cans tunafish
2 18 oz. Tang bottles	2 18 oz. Tang bottles

Now we ask how much each of these costs at each store. For example, here is the computation of the Milk and Eggs basket cost at Mr. G.

$$
\begin{aligned}
3 \times .58 &= \$1.74 \\
2 \times .65 &= 1.30 \\
1 \times .79 &= .79 \\
0 \times .37 &= 0.00 \\
2 \times .98 &= 1.96 \\
\text{Total} &\quad \$5.79
\end{aligned}
$$

Another example might be that of the Meat and Fish basket at Pete's.

0 × .67 =	$0.00
0 × .65 =	0.00
4 × .79 =	3.16
2 × .43 =	.86
2 × .98 =	1.96
Total	$5.98

After carrying out all eight of these computations, we have the following results. (You should check them.)

	Mr. G	A&P	Coop	Pete's
Milk and Eggs	$5.79	5.86	5.48	6.06
Meat and Fish	5.86	6.34	5.96	5.98

Notice that the prices of the baskets vary quite a bit among the stores: from $5.48 to $6.06 for the Milk and Eggs basket, and from $5.86 to $6.34 for the Meat and Fish basket.

What's more, the relative standings of the stores depend on the basket chosen. Let's list the stores in order of price level.

	Least expensive		Most expensive	
Milk and Eggs	Coop	Mr. G	A&P	Pete's
Meat and Fish	Mr. G	Coop	Pete's	A&P

You can see that the positions of the stores in the price orderings depend on the basket chosen. For the two above baskets, however, the Coop and Mr. G are the two less expensive stores in both cases, and the A&P and Pete's are the two more expensive.

Times, prices, and stores change. These calculations apply only for one moment in 1969. Things may be very different now.

Exercises

1. Try your hand with some other baskets. Find one that moves Mr. G into the next most expensive position. [Hint: Try a basket with lots of Tang.]

2. For no basket will Mr. G turn out to be more expensive than Pete's. Explain why by comparing the columns of prices for those two stores.

Next we look at Table 1, the full table of results from the neighborhood newspaper survey.

The full survey covered many more items than those in the extract, and nine grocery stores were covered,

Table 1. Grocery survey

Units for market basket	Product	Mr. G	A&P	Del Farm	National	Coop	Morgan	Food Mart	Grocer-land	Pete's
3	Milk, 1/2 gal. 2%	.58	.59	.59	.58	.57	.62	.61	.56	.67
1	Cottage Cheese, 1 lb.	.41	.40	.42(a)	.40	.39	.44	.44	.39	.46
1	Margarine, Blue Bonnet, 1 lb.	.29	.33	.32	.29	.32(a)	.32(a)	.32(a)	.32(a)	.35
2	Eggs - 1 doz. large	.65	.69	.55	.69	.57	.69	.75	.69	.65
1	Philadelphia Cream Cheese, 8 oz.	.39	.33	.39(a)	.39	.39	.39	.41	.41	.43
1	Oscar Meyer Hot Dogs, 1 lb.	.79	.95	.89	.89	.85	.79	.83	.98	.79
1	Flour - Gold Medal, 5 lbs.	.55	.59	.59	.59	.59	.61	.69	.69	.69
1	Oscar Meyer Liver Sausage, 8 oz.	.39	.49	.59	.49	.47	.45	.45	.45	.47
1	Brown Sugar, Domino, 1 box	.175	.19	.17	.175	.175	.185	.19	.19	.19
1	Quaker Oats - 18 oz.	.31	.29	.30	.31	.35	.33	.33	.33	.33
3	Bread - 20 oz., Silvercup	.33	.33	.33	.33	.33	.33	.33	.33	.33
1	Noodles Almondine - Betty Crocker	.45	.49	.33	.46(a)	.49	.47	.47	.49	.49
1	Joy Detergent - 22 oz.	.57	.57	.59	.57	.57	.63	.65	.69	.82
2	Scott Toilet Paper, roll	.145	.145	.14	.15	.15	.165	.156	.156	.165
1	Coffee, Hills Bros., 1 lb.	.80	.81	.82	.81	.81	.845	.89	.89	.93
1	Heinz Catchup, 14 oz.	.24	.26	.21	.24	.29	.30	.29	.29	.29
2	Tunafish, Chicken of Sea, 6½ oz.	.37	.39	.40(a)	.36	.39	.41	.45	.40(a)	.43
1	Strawberry preserves, Kraft, 8 oz.	.39	.45	.39	.39	.39	.40(a)	.40(a)	.40(a)	.43
12	Baby Food - Gerber, Beech Nut	.11	.11	.11	.11	.10	.115	.123	.123	.13
2	V-8 Juice, 12 oz.	.145	.145	.15	.145	.145	.15	.15	.15	.15
2	Hunts Tomato Sauce, 8 oz.	.11	.145	.13	.11	.13	.13	.13	.13	.165
1	Skippy Peanut Butter, 12 oz.	.41	.45	.41	.43	.45	.47	.47	.49	.49
2	Campbells Split Pea Soup	.18	.195	.17	.183	.185	.21	.196	.19(a)	.195
1	Corn, Green Giant, 12 oz.	.25	.24	.27	.23	.265	.27	.275	.26(a)	.285
1	Miracle Whip, 1 pint	.39	.41	.45	.37	.45	.45	.45	.49	.49
2	Campbells Pork & Beans, 1 lb.	.15	.185	.17(a)	.15	.175	.18	.175	.175	.175
2	Chef Boyardee Ravioli	.31	.35	.35	.29	.35	.35	.35	.37	.39
2	Mary Kitchen Roast Beef Hash	.59	.57	.49	.48	.57	.59	.59	.59	.59
1	Tang, 18 oz.	.98	.88	.85	.88	.89	.85	.97	.98	.98
1	Ritz Crackers, 1 lb.	.47	.47	.47	.47	.47	.47	.49	.47	.49
1	Oreo Cookies, 1 lb.	.51	.53	.53	.49	.53	.53	.53	.53	.59
1	Carnation Evaporated Milk, 1 can	.18	.19	.18	.18	.195	.196	.22	.205	.21
2	Kraft Macaroni & Cheese, 7¼ oz.	.195	.21	.19	.195	.225	.23	.245	.215	.245
1	Cheerios, 10 oz.	.35	.37	.37	.35	.39	.41	.39	.39	.38
1	A-1 Sauce, 5 oz.	.34	.34	.39	.36	.39	.39	.39	.39	.39
1	Karo Corn Syrup	.29	.35	.35	.35	.35	.37	.39	.39	.39
1	Noodles Romanoff, Stouffers frozen	.49	.59	.51(a)	.51(a)	.55	.49	.49	.49	.49
	Market basket cost	$20.16	21.10	20.35	20.20	20.71	21.51	22.12	21.89	22.73

Note (a): If an item is not carried by a store, it is assigned the average price from the stores that do carry it.

Source: Hyde Park-Kenwood Voices, iv, No. 9 (September 1969): 10. Reprinted by permission.

not just the four that we started with. Let's first discuss some aspects of the full table that may be confusing.

a) Tenths of cents. Some prices are given to tenths of a cent; for example, the price for brown sugar at Mr. G is .175, that is, 17½ cents. Presumably the reason for this is that the sugar was priced at 35 cents for two boxes. Another example is the price for baby food at the Harper Sq. Food Mart: .123, or 12.3 cents. Presumably this means that baby food was priced at 37 cents for three jars, or 12 1/3 cents a jar. In decimals this is 12.333... cents, and it is reasonable to round it off to 12.3 cents or .123 in dollar units.

A case that is not quite so clear is the price for a can of Campbell's Split Pea Soup at National: .183. One plausible explanation would be that cans of that soup were sold at a price of 55 cents for three cans; dividing 55 by 3 gives 18.333..., and rounding leads to 18.3 cents. It is also possible that .183 represents an error in computation or a typographical error, problems to which we are all subject and ones that can occur even after repeated checks.

b) Missing items. In some cases, a given grocery item was simply not carried by a particular store at the time of the survey; for example, Blue Bonnet margarine was not carried by four of the nine stores. In such cases, the table could have had just a blank space, but then we would be puzzled about how to compute comparative prices for market baskets that include the sometimes-missing item. In this survey, it was decided to fill in the blanks by using the average price for all stores that did carry the item.

For example, in the case of margarine the prices for the five stores that carried it were

.29, .33, .32, .29, .35.

Add these up to get $1.58, and divide by 5. We got .316. This was then rounded to .32, and the resulting average price was used for the stores that did not carry margarine.

For more on missing prices, see the end of this set.

Exercise

3. Try your hand at computing some of the other averages for items with missing prices, and see whether the missing prices were correctly filled

in. (There is at least one small error, around the middle of the table. Can you find it?)

c) The big market basket. The numbers in the left-most column of Table 1 are the numbers of units used in the survey for defining the market basket, that is, for providing a basis for comparing the nine stores in just the same way that we looked at several small baskets for the extract at the beginning of this discussion. How this particular market basket was chosen is not discussed in the text of the articles going with the table, but probably it was based on guesses about how often the separate items were purchased.

The numbers in the bottom row of the table show the market-basket costs for all nine stores; they range from $20.16 to $22.73, and they are found in just the same way as in our earlier discussion. For example, the $20.16 for Mr. G was found by starting with

$$3 \times .58 = 1.74$$
$$1 \times .41 = .41$$
$$1 \times .29 = .29$$
$$2 \times .65 = 1.30$$

Exercise

4. Complete this computation and check that the answer is indeed $20.16 as listed in the bottom row. Similarly, check the market basket costs for one other store. Spread the stores around the class so that the computations for all stores get checked.

d) Quality. In surveys of this kind, it is important to be sure that items of the same name in different stores are really of the same quality. That is why nearly all the items in Table 1 are name brand items that are sufficiently standard so that quality variation is not a cause of concern. The text going with the table says that "fresh meat was ... left off the market basket because wide variations in quality and fluctuations in price make surveying unreliable."

This quotation raises a second point too. It was probably impossible to survey all nine stores in one day, or even in two or three days. That makes problems if the market basket includes items with prices that vary from day to day. For example, suppose one store a day is surveyed and that fresh meat prices tend to dip on Fridays and Saturdays. Then the stores that happened to be surveyed on those two days would, if meat were in the basket, have an unfair advantage in the price competition.

In Table 1, three items do not have brand names: milk, cottage cheese, and eggs. (Milk and eggs, however, have brief descriptions, "2%" and "large", respectively.) Whether or not this is a serious matter depends on how much these items vary, for example, in freshness. In large, careful price surveys, like those done by the government, the specification of quality requires a lot of hard, professional work. There is always a need for balance, since the more narrowly quality is defined, the more missing items there will be.

Other aspects of quality refer to an entire store, for example, stores have different policies regarding delivery, parking, charge accounts, hours, etc. These are difficult to take into account.

Project

Organize and carry out a small price survey of some grocery stores in your community. Keep it modest in size. You will want to think about such matters as

Which stores to choose. (We recommend four to six.)
Which foods to choose. (Do not pick very many.)
What units and brand or quality designations to use.
When to make the observations.
What to do about missing items.

It would be a good idea to have two or more observers obtain prices from each store separately. Then you can compare the prices for each store and try to resolve any differences.

After obtaining the prices, think about reasonable market baskets for the item you chose. Compare the price levels of the stores for two different baskets.

If possible, repeat the survey after a month or two, and see whether the relative price levels of the stores have changed.

More on missing prices. (This section is a little more advanced and can be omitted without loss of continuity.) The above method of filling in missing prices has the advantage of simplicity. It can, however, lead to a misimpression. Suppose, for example, that there are two stores, A and B, and that (unknown to the survey people) store B consistently has prices 50% higher than those of store A. Suppose further that there are only two items in the market basket, and that the results from a survey are

Item	Store A	Store B
1	.60	.90
2	.40	missing

Using the average of the other stores to fill in a missing price amounts here to using the price .40 for item 2 in store B, whereas the price would have been .60 if the item had been carried. Thus we have two tables.

Survey table

Item	A	B
1	.60	.90
2	.40	.40(a)

"True" table

Item	A	B
1	.60	.90
2	.40	.60

If the market basket used for comparison of the stores had one unit of item 1 and three units of item 2, the basket costs would be

Survey market basket prices

A	B
$1.80	$2.10

"True" market basket prices

A	B
$1.80	$2.70

So the comparison is appreciably affected.

This problem might be a troublesome one in a grocery store survey since supermarkets are less likely to have missing items than smaller food stores and supermarket prices tend to be lower than prices in smaller stores. Hence the "average" method of filling in prices will make differences in market basket costs <u>less</u> extreme, as between supermarkets and smaller stores, than they would be if all stores carried all items.

There are better ways of filling in prices, ways that use information about relative prices in stores for other items. But these methods are too lengthy for discussion here. A usable approach of a completely different kind is to ask the store manager what his price for a missing item would be if he carried it, or what the price was the last time he carried it. (Of course this requires a cooperative and honest manager, with a good memory.) Another practical approach is to substitute for a missing item the closest substitute item that the store carries. Can you think of some dangers or problems with this approach?

Turning the Tables

JOEL E. COHEN

INTRODUCTION

Sometimes something happens which seems so unusual that we wonder what makes it happen that way. The real question is, what makes us think it is unusual? Most people think something is unusual if it is uncommon. The purpose of this example is to try to make you see how we can count possibilities to help determine whether a given event is really common or uncommon, and thus whether we should think of it as ordinary or as so unusual that it demands some special explanation.

The student cafeteria of a university in California has small square tables and rectangular tables. Some students come into the cafeteria between meals to use the tables for casual conversation and joint studying. A psychologist interested in how people occupy space observed the students for a period of months.

He concluded that 50 pairs seated at the small square tables showed a preference for corner rather than opposite seating because 35 pairs sat corner-to-corner while 15 sat across from one another. He inferred a similar preference for corner seating from a subsequent study of eating pairs in a hospital cafeteria; here, of the 41 pairs he observed, 29 sat corner-to-corner while 12 pairs sat across from one another.

After the rectangular tables in the student cafeteria were modified, but the square tables were left

The author is at Harvard University, Cambridge, Massachusetts.

the same, the psychologist observed for several more months.

He distinguished between those pairs who were interacting (conversing, studying together) and those who were "coacting" (occupying the same table but studying separately). Of the 124 pairs he saw seated at the small square tables, 106 were conversing or otherwise interacting while 18 were coacting. Again he asserted that the interacting pairs showed a definite preference for corner seating, because 70 were corner-to-corner while 36 were seated across from one another. However, coacting pairs chose a very different arrangement. Only two pairs sat corner-to-corner and the rest sat opposite one another. These results, the psychologist claimed, supported the previous inference that corner seating was preferred over opposite seating in a variety of conditions where individuals interact. To the psychologist, the observations suggested that corner seating preserves the closeness between individuals and also enables people to avoid eye contact, since they do not sit face to face.

Do the psychologist's data actually show that people prefer corner to opposite seating at small square tables if they are interacting? Do the data support his claim that corner seating is preferred because it enables people to avoid eye contact? What alternative explanations, consistent with his observations, are possible?

For convenience, summarize the data:

	Number of pairs observed	
Interacting pairs	Corner	Opposite
Student cafeteria, first series	35	15
Hospital cafeteria	29	12
Student cafeteria, second series	70	36
Total	134	63
Coacting (not interacting) pairs	2	16

Obviously many more pairs chose corner seating than opposite seating. But before we believe that this shows students prefer corner seating, we need to know what to expect if students show no preferences at all about their seating.

The possible arrangements of two students (×) at a square table are:

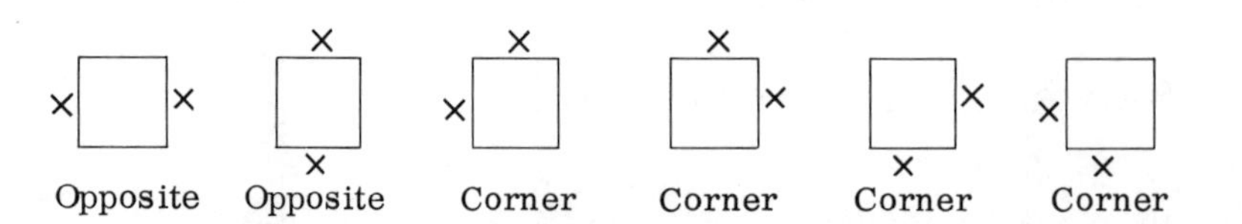

There are twice as many corner as opposite arrangements. If students chose at random among these possibilities, then twice as many corner seatings as opposite seatings would be expected.

Another way of seeing the same result is to suppose that the first student of the pair chooses any seat. For the second student, the two seats to the right or the left of the first give corner seatings; only one seat is opposite. If the second student has no preference, then twice as many corner as opposite seatings will occur.

Since the psychologist observed a total of 134 + 63 = 197 interacting pairs, he should have expected approximately 2/3 × 197 = 131.3 corner seatings and approximately 1/3 × 197 = 65.7 opposite seatings if interacting students showed no seating preferences. The observed data are so close to these expectations that they do not support the conclusion that interacting students prefer corner seating.

Since 18 coacting (not interacting) pairs were observed, 12 corner seatings and 6 opposite seatings would be expected if students showed no preference. Only 2 corner but 16 opposite pairs were observed. Hence it is probably safe to conclude that people who avoid interacting prefer not to sit at a corner with each other.

Even though the proportions in the data agree with those predicted by random seating, we haven't proved the randomness. Some pairs may prefer, or have the habit of, sitting one way, and some another. This could be checked by watching the same couples on several days. Beyond this, it is possible that they sit down at random the first time and then do the same thing thereafter.

What do you think that asking the people in the cafeteria about their preferences would add to the investigation? What new problems of interpretation arise from such interviews?

Exercises

1. Suppose the square tables were placed with one edge against the wall. How many different seating arrangements are possible for a pair of students? How many of these arrangements are opposite and how many corner-to-corner? Does it matter that the psychologist didn't say whether or not the square tables had one edge against the wall?

2. Suppose the cafeteria had circular tables with six chairs each. If a team of four students sat down randomly at one of these tables, what is the chance that the two empty seats would be next to

each other? What is the chance of an arrangement with one student seated between the two empty places? What is the chance that the two empty seats would be opposite each other? From these results, prove that empty chairs usually do not face each other.

3. Tricky question: Assuming four students sit down randomly at a six-seated round table, what is the chance that a student will have empty places on both sides of him? (Remember that in the seating arrangement with one student seated between the two empty places, only one of the four students has empty places on both sides of him. Counting the arrangements from the point of view of an individual student may be different from counting the arrangements of the table as a whole.)

Testing Beer Tasters

WILLIAM H. KRUSKAL

INTRODUCTION

This example is very different from the previous ones, yet its purpose is much the same. How do you count the ways something can happen to help decide whether an outcome is usual, or so unusual as to be worth comment and explanation?

> A magazine advertisement in 1967 began as follows:
>
> WILL THE REAL MILLER HIGH LIFE
> DRAFT BEER PLEASE STEP FORWARD
>
> We've tried our famous Miller High Life taste test on many a "beer expert"; pouring three glasses, one from the tap, one from the can, and one from the bottle. And then we've asked which is which.
>
> Result? No one, up to now, has identified the three correctly. Why?
>
> All three glasses have the same distinctive Miller High Life flavor.*

The advertisement claims that beer experts cannot tell the taste differences between Miller High Life beer on

The author is at the University of Chicago, Chicago, Illinois.

*An advertisement of the Miller Brewing Company in The New Yorker, 22 April 1967, p. 69. Reprinted by permission.

tap, from a can, and from a bottle. The evidence given is that not one beer expert out of many has correctly identified all the three glasses.

Exercise

1. Before reading the discussion, think critically about the claim and the evidence. Ask yourself whether anything seems strange.

Discussion. Something strange was indeed brewed by the Miller High Life advertising agency. Suppose, for the sake of discussion, that the three ways of packaging the beer really do lead to indistinguishable glasses. Then each way of assigning, or guessing, which glasses are which should be equally probable. In other words, suppose that the guessing is random.

There are exactly six ways or patterns for saying which glass is which. Each row in the tabulation below shows such a pattern.

Glass number		
1	2	3
T	C	B
T	B	C
C	T	B
C	B	T
B	T	C
B	C	T

Here T, C, and B stand for Tap, Can, and Bottle respectively.

In any particular trial, exactly one of these six patterns would be correct and the others wrong. Hence, if the three kinds of beer are indistinguishable, then the six ways of assigning labels are equally likely, and the chance that a single expert is wrong is 5/6. The chance that two experts are both wrong would be $(5/6) \times (5/6)$, assuming that they do not consult, or otherwise affect each other's choice. Similarly, the chance that all the n experts are wrong would be $(5/6)^n$. Note that "wrong" means that there is at least one error (indeed, at least two). An expert might guess the tap beer glass correctly but mix up the other two; for present purposes he would be counted as wrong.

How big is n? It depends on what the ad writer meant by "many", and the advertisement does not say. On the next page is a table that will be helpful.

For example, if there were 10 beer experts, then the chances are about 4 in 25 (that is, 0.16) that none would make all the assignments correctly if the three glasses were indistinguishable. But if there were 30 experts, the probability that none would make all correct assignments goes down to less than 5 out of 1000, quite a small figure. If there were 50 experts, the probability is only a bit more than one out of 10,000.

n	Chance that all n experts are wrong, $(5/6)^n$, to two significant figures
1	5/6 = .83
2	25/36 = .69
5	.40
10	.16
20	.026
30	.0042
40	.00068
50	.00011

Exercise

2. Check the above table for n = 10, 30, and 50. Use logarithms.

So it seems reasonable to think that perhaps there were not so many beer experts after all. That would explain the reported strange result. There are other possible explanations, but before going to them, here is a project that will help you understand the above reasoning.

Project 1

This requires ordinary dice, preferably one for each member of the class. Suppose that the six sides of a die correspond to the six possible patterns in the tabulation at the start of the Discussion, numbering the patterns from the top down. Suppose, to be specific, that pattern 1 (T,C,B) is the correct one, and that in fact the three kinds of beer are indistinguishable. Then each pattern has probability 1/6, just as each face of a well-made die has probability 1/6.

Each member of the class should toss a die 10 times and record whether a 1 appears or fails to appear sometime during the 10 tosses. The tosses should be vigorous and the die should bounce against some hard vertical surface, like an upright book.

If the tosses correspond to theory, then about .16 of the class members should find no 1's among their tosses. For example, if there are 30 students in the class, about 30 × .16, or roughly 5, should find no 1's. You might well not come out with exactly 5, but for a class of about 30, the number who fail to get 1's should be between 3 and 7 more than half the time.

If your class is a small one, each member should go through the sequence of tosses two or more times, so that there are at least 30 sequences in all.

Other explanations. Now let's talk about other possible explanations for the strange tasting results. Before reading the following paragraphs, try and think of other explanations yourself.

Systematically wrong assignments. One alternative explanation is that the three ways of packaging beer are really distinguishable, but in ways that lead systematically to wrong assignments. It might well be that the experts have expectations about the taste of beer when it comes from a tap, from a bottle, or from a can, but that somehow these expectations are simply wrong, at least for Miller High Life beer. For example, perhaps the canned beer tastes to experts like bottled beer, the bottled like tap, and the tap like canned.

If the individual assignments of labels by the experts were available, one could investigate this possible explanation. For example, if the correct pattern were T,C,B and most of the experts assigned the pattern C,B,T, then this kind of systematic error might seem plausible.

Similarly, if this problem had been thought over earlier, it might have been solved by a training session before the trials themselves. In the training session, the experts would have been given identified samples of the three kinds of beer.

Identification label effects. Another alternative explanation for the strange results might be that apparently neutral identification of the glasses had the effect of introducing systematic error. Suppose that the glasses of bottled beer were labeled "A", the glasses of canned beer "B", and the tap beer glasses "C". Suppose further that each taster was given a slip like this to fill out, with instructions to put the appropriate letter in each box.

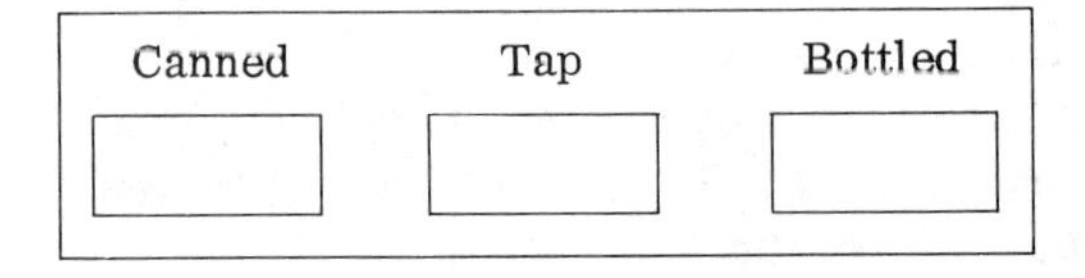

The correct entries are B, C, and A in order, but it might be that the order A, B, C (a wrong ordering) is psychologically preferred because of our familiarity with the alphabet. To be specific, suppose that the wrong ordering A, B, C has probability 0.5, and that each of the other orderings has probability 0.1.

Exercise

3. Show that under these circumstances the probability that none of the n tasters hit the correct ordering is $(9/10)^n$.

Of course $(9/10)^n$ is larger than $(5/6)^n$.

As before, examination of the detailed results might well tell us something about such identification label effects.

If this problem had been thought of beforehand, it might have been taken care of by varying the identification label scheme from taster to taster, or by varying the order of the three words on the answer slip.

Dependence effects. Until now, we have assumed that there are no influences from taster to taster--each makes his guesses independently of the other tasters. Were this not so, the observed absence of the correct pattern might result from the dependence. To take an extreme example, suppose that the first expert to sip from the glasses was a man of high reputation as a beer expert, and suppose that his guesses were announced to the others before they began their tasting. Suppose further that the physical arrangement of the tasting set-up was such that later tasters could know which glass corresponded to each glass of the first taster. Then there might be a strong tendency for the later tasters to be copycats and to follow the pattern set by the first one. If his pattern was wrong, the others would also tend to be wrong. In any case, the simple computations given at the beginning of the discussion would be inapplicable.

There might be other, more subtle, sources of dependence among the tasters. Facial expressions, sounds of pleasure or dislike, and small body motions can all convey a good bit of information, as experienced card players know.

Still other explanations. We have discussed several alternative explanations. There are others; for example, everyone might have been so relaxed after a few glasses of beer that the record-keeping became sloppy. As we have seen, in order to discuss the alternatives in detail, we would have to know a lot more about how the testing was done, and about the individual patterns of assignment. But an advertisement in *The New Yorker* is not an article in a scientific journal, and even articles in scientific journals may omit important background information and details of data.

Applications. A beer advertisement in a magazine is clearly not a matter of great importance, except to the beer company and its advertising agency. Questions like those we have discussed, however, arise in other situations. For example, industrial concerns making foods of all kinds use panels of tasters to evaluate routine production and to measure the tastiness of experimental new products. Large organizations use tasters before making huge purchases of food products; for example, the U.S. Army has an elaborate system of coffee tasting. Consumer organizations use tasting panels in their testing.

Details differ from case to case: there may be four or six variants of a given product, not three; the tasters may be experts or not; the product may not be food or drink, but perfumes, clothing, razors, etc. Nonetheless, the considerations we have discussed above and others, are seriously thought over in evaluation work of this kind.

Project 2

Take three presumably identical bottles of a soft drink, pour each into a pitcher, and secretly mark the pitchers T, C, and B. Ask your classmates to try little glasses from the pitchers, and to suppose that one comes from a soda-fountain tap, another from cans, and the third from bottles. Ask them to guess which is which.

It would be a good idea to use small slips like the one illustrated earlier, but to have them made up in all the six possible patterns and to mix up the six kinds. Then you could put clearly visible labels A, B, and C on the pitchers and ask the tasters to fill in letters on the slips according to their guesses. The tasting should be done behind a screen or in a separate room, and you should arrange matters so that a student who has completed the tasting does not go back to the group waiting their turns. This will help to avoid dependence.

Do not assume that it is a simple matter to set up even an uncomplicated project like this. It will take careful planning. You might well have one class arrange the project and ask another class to do the tasting.

After all the tasters have made their assignments, tabulate the results and see how many have made at least one wrong assignment. Compare this with $(5/6)^n$, where n is the number of tasters.

Carry the tabulation further to see how many have no errors, exactly one error, exactly two errors, and exactly three errors. Compute the corresponding probabilities under the assumption that the guesses are random. First think about the probability of exactly two correct assignments out of three; the answer is zero.

Multiply each probability by n and compare these theoretical results with the observed numbers.

Project 3

Now go through the same procedure but with three soft drinks that are really different, yet not very much different. For example, take three competitive cola drinks.

You might want to have a training session before the actual test tasting.

If the drinks *look* different, it might be a good idea to blindfold the tasters before they do the tasting. Without blindfolds, you would be examining ability to discriminate among the drinks by using both taste and vision; with blindfolds, you would be examining only taste.

Compare the results from Projects 2 and 3.

Estimating the Size of Wildlife Populations

SAMPRIT CHATTERJEE

INTRODUCTION

How do you make guesses for the size of a population when you can't count its members exactly? By studying a simple problem involving fish in a small pond, we hope you will see how a much larger and more realistic problem, such as a ranger's estimate of the number of deer in Yellowstone Park, might be handled.

If just guessing isn't good enough, how would you estimate the number of fish in a lake? Before reading further, try to think of some methods. Keep in mind problems of money, time, and the fate of the fish; for example, you might not want or be able to catch them all.

A method often used in problems of this kind is called the capture-recapture method. Suppose we want to estimate the number of fish in a lake. The idea is to catch some of the fish, count them, tag them, and return them to the lake. Later, we again catch some fish, count them, and also count the number of recaptured fish, that is, the number that had earlier been tagged. We then use these three numbers to make an estimate of the total number of fish in the lake.

There are certain problems about procedure. How do we pick the spot where we fish? Should we go back on the second day to the same place or places where we fished before? Another problem is: are the fish caught the first time hurt by catching and tagging, so that some die? We answer this by being careful in

The author is at New York University, New York, New York.

handling them and returning them quickly to the lake. Now let's be more precise in our description of the method as applied to fish in a lake.

Spots in the lake are chosen, and a catch of fish is made. The fish in the catch are counted, tagged, and quickly released alive into the lake again. Soon, but not too soon, after, a set of spots in the lake is again chosen, and a second catch is made. In the second catch, the total number of fish caught is counted, as well as the number of fish that had been tagged before.

Start out by assuming that we have a very small pond and only a few fish. Suppose our first catch is 10 fish, which we tag and throw back. Our next catch is 7 fish, of which we find 2 tagged. Now we try to estimate how many fish there are in the pond. The first thing we can decide is the smallest possible number of fish which could be in this pond. What is it? How many fish did we tag? 10. How many untagged fish did we catch on the second day? 5.

Therefore, there must be at least 10 + 5 fish in the pond. But if there were only 15 fish, 10 tagged and 5 untagged, it does not seem very likely that when we caught 7 on the second day we would manage to get all 5 of the untagged fish and only 2 of the tagged ones. So we guess there are probably more than 15, but the problem is how many more. What would be a reasonable guess? If there were 1000 fish in the lake, 10 tagged and 990 untagged, does it seem likely we would have caught 2 tagged fish? Most people would agree that it does not. Neither 15 nor 1000 seem like good estimates. What is a better estimate? Of course, we have to assume that the tagged and untagged fish are all mixed up in the pond and that catching a fish once does not either scare it away or make it more easily caught a second time.

In these circumstances, most people feel that since we caught 2 tagged fish out of 7, probably about 2/7 of the fish in the lake are tagged. Since we tagged 10 fish and 10 is 2/7 of 35, there are probably about 35 fish in the pond. There could be more, many more, or fewer (but not fewer than 15), but 35 is an estimate many people would be comfortable with.

As another illustration, suppose we have a bag of marbles with too many in it for convenient counting, but still not so many that good mixing is impractical. Take a handful of marbles, and suppose you have picked up 8. Put a mark on each of the eight, replace them in the bag, and shake it thoroughly to mix the marked marbles with the unmarked ones. Now take out another handful and find 20 marbles of which only 1 is marked. What is a reasonable estimate of the number of marbles,

n, in the bag? The smallest possible number of course is 8 + 19 = 27, but we guess there are many more than that. What would a reasonable guess be? We might assume that the fraction of the marked marbles on the second draw is approximately the same as the fraction of the first draw is of the total number of marbles in the bag. This leads to the equation

$$\frac{1}{20} = \frac{8}{n},$$

which gives us $n = 160$, and that is the estimate we make.

To see the general picture we must use some mathematical notation. In the first example above,

let n_1 be the number of fish in the first catch. All of them are tagged and returned to the lake.

Let n_2 be the total number of fish in the second catch.

Let t be the number of tagged fish in the second catch, and

let n be the total number of fish in the lake.

Then n is the number we are trying to estimate, using the three observed numbers n_1, n_2, and t. We know that n is certainly greater than $n_1 + (n_2 - t)$, and a reasonable estimate for n seems to be that one for which the proportion

$$\frac{t}{n_2} = \frac{n_1}{n}$$

holds, that is, the proportion of tagged fish caught on the second day is the same as the proportion of tagged fish in the lake. This gives us

$$n = \frac{n_1 n_2}{t}, \quad t \neq 0.$$

Elsewhere, we show [2] that this value of n is the one for which the chance of catching t tagged fish in a total catch of n_2 fish is the highest possible.

Now that we have an estimation method, let us think about the assumptions that we made. Try to think about this problem before reading on. Some of the assumptions that must hold are:

a) the fish tagged are not affected by tagging, and the tags will not come off;

b) the tagged fish become completely mixed in the population;

c) the fish in the lake are equally available for capture whatever their positions in the lake;

d) the population does not change between the two catches, for example, additional fish are not introduced from a hatchery, and the lake does not overflow, allowing the fish to escape;

e) the effect of birth and death are negligible between catches. To ensure this, the second catch should be taken soon after the first catch.

The first two assumptions depend on the season and the species of fish, but fishery experts maintain they are approximately true and can be assumed without too much error. We strengthen the reality of the third assumption by drawing the spots to be fished at random.

Be sure to do Exercise 1 about estimating seal populations.

Exercises

1. The data given below represent the results of a capture-recapture experiment conducted in Alaska in 1961 to estimate the fur seal pup population. The results come from St. Paul Island which has 12 fur seal rookeries. The total number of fur seal pups on the island is obtained by adding up the estimates for each of the rookeries. By observation it has been found that there is little or no movement of the pups between the rookeries. We present the data for only one of the rookeries on the island, named Gorbatch. The pups were marked by shearing the black guard hair from the top of the head to expose the light underfur. This method produced an easily identified mark that could be seen without difficulty. Formerly this was done by using a cattle-car tag and an identifying checkmark on the flippers. In Gorbatch rookery, 4965 fur seal pups were tagged in early August. In late August, in a sample of 900 fur seal pups, it was found that 218 of them were previously tagged. Show that the estimate of fur seal pup population in Gorbatch rookery in 1961 was 20,500 (correct to the nearest thousand). More details about the study can be found in a paper by D. G. Chapman and A. M. Johnson entitled, "Estimation of Fur Seal Pup Populations by Randomized Sampling," printed in the Transactions of the American Fisheries Society, 97, No. 3, (19 July 1968): 264-270.

2. From a deck of cards, remove a pile of cards. Use the capture-recapture method to estimate how many cards are in the pile you take off. Deal off a certain number of cards, and record the cards dealt. Return them to the pile and shuffle the cards in the pile thoroughly. (Trying to shuffle thoroughly is the difficult part of this exercise.) Now deal off some cards, count the previously recorded ones and estimate the number in the pile. Do this several times, get the average of all your estimates, and finally compare with the actual count of cards in the pile.

3. Rangers in a park caught 100 deer, marked them, and released them. Shortly afterward they caught 200 deer and found 5 marked ones. What is an estimate of the number of deer in the park? Do you think you have a reasonable estimate of the number of deer in the park?

4. The data are identical with that of Exercise 3 except that the survey was done the following year. Comment on the estimate and the possible changes that might be needed.

5. Suppose the same counts as in Exercise 3 were made in a state where hunting is allowed. Would your formula for estimating need to be reexamined? Why or why not?

6. Are estimates of wildlife populations made this way likely to be more or less accurate than estimates of such things as the number of marbles in a bag? Why or why not?

7. Estimating the size of a wildlife population is fascinating, difficult, and useful. As an introduction to it, try to devise methods for one of the following (each offers special problems, and you may need to read up on the habits of these populations):

   a) sea gull population on the Eastern seaboard of U.S.A.

   b) the seal population of Alaska

   c) the polar bear population in the world

   d) the number of lions in Kenya

   e) the number of tigers in India.

References

[1] You may enjoy reading the article on "Estimating whale populations" by Douglas Chapman in Judith

Tanur *et al*. (Editors), *Statistics: A Guide to the Unknown*, Holden-Day, San Francisco, California, 1972.

[2] *SBE*, *Finding Models*, Set 3.

Tom Paine and Social Security

WILLIAM H. KRUSKAL
AND RICHARD S. PIETERS

INTRODUCTION

This example shows how one author made serious errors in estimating the sizes of groups in the population. The author is Tom Paine, who lived at the time of the American Revolution. He tried to estimate the number of old people in the population. More recently others have made mistakes similar to his with much less excuse, since good statistical methods and good population data are now available for this kind of work. Our aim is to make you see how difficult such estimation is, how easy it is to make mistakes, and how to prevent them.

Most of you know something about Thomas Paine, the English political thinker and writer who came to America in 1774 and put his skills to use in opposing British rule of the American colonies. Paine supported independence of the colonies in his famous pamphlet _Common Sense_, and a little later he published a series of pamphlets, called _The American Crisis_, in eloquent support of the revolution. The first issue of _The American Crisis_ (19 December 1776) began with the following ringing words.

> "These are the times that try men's souls: The summer soldier and the sunshine patriot will, in

---

This example was brought to the authors' attention by Robert Streeter, University of Chicago. William H. Kruskal is at the University of Chicago, Chicago, Illinois; and Richard S. Pieters is at Phillips Academy, Andover, Massachusetts.

> this crisis, shrink from the service of his country; but he that stands it NOW, deserves the love and thanks of man and woman. Tyranny, like hell, is not easily conquered; yet we have this consolation with us, that the harder the conflict, the more glorious the triumph. What we obtain too cheap, we esteem too lightly..."

After the American revolution, Paine returned to England where, in 1791 and 1792, he published The Rights of Man. This important book opposed the monarchical system of England, and defended the principles of the French revolution that had recently taken place. The British government, not surprisingly, regarded The Rights of Man as inflammatory and seditious; Paine avoided arrest and trial by fleeing to France.

Our statistical example begins with a quotation from The Rights of Man. Paine presents arguments, based on quantitative reasoning, for universal education, government financial help for the aged, and other such national welfare activities. In particular, Paine requires an estimate of the number of old people, so that he can have a rough idea of the cost of providing them with financial aid. Modern industrial societies generally use an age of about 65 for defining elderliness. But for his survey Paine settles on 50. The quotation follows.

> "To form some judgment of the number of those above fifty years of age, I have several times counted the persons I met in the streets of London, men, women, and children, and have generally found that the average is about one in sixteen or seventeen [who are older than fifty]. If it be said that aged persons do not come much in the streets, so neither do infants; and a great proportion of grown children are in schools and in workshops as apprentices. Taking, then, sixteen for a divisor, the whole number of persons in England of fifty years and upwards, of both sexes, rich and poor, will be four hundred and twenty thousand."

Exercises

1. How many people lived in England at this time if Paine was right in saying that the 420,000 old people were one sixteenth of the population?
2. What is the percentage of "people fifty years and upwards" corresponding to Paine's "one in sixteen"? What percentage corresponds to "one in seventeen"?

3. Before reading on, think about Paine's procedure for estimating the proportion of people 50 and over. What difficulties do you see?

Discussion of Exercise 3. There are a number of rather obvious difficulties with Paine's procedure. How many of the following did you list?

a) Paine at one point speaks of people "above fifty years of age", but later he speaks of "fifty years or upwards". This is a minor ambiguity, but it might make us wonder about the care with which Paine carried out the survey.

b) Another minor ambiguity is that Paine says "one in sixteen or seventeen", but then somehow settles on sixteen.

c) More important is the difficulty of telling from a glance on the street whether a person is 50 or over. There must have been many people in their 40's and 50's of whom Paine could not be sure.

d) Paine walked in the streets of London, but the streets on which he walked may not have been typical in terms of age distribution. For that matter, the age distribution in London may have been different from that of the whole country.

e) There is absolutely no reason to think that people 50 or over stay off the streets as much as "infants" or "grown children".

Paine might well reply that the above are quibbles, and that for his purpose even a very rough estimate was satisfactory. Nonetheless, in planning new social welfare programs it is of great importance to get a reasonably good idea of costs beforehand.

There is one further objection to Paine's procedure that could affect his figures considerably and that applies even if all the above objections can be anwered. It relates to the different amounts of time that people of various ages spend on the streets.

Exercise

4. Can you think of this further objection?

Discussion of Exercise 4. Paine is tripped by what may seem a reasonable argument into an interesting mistake. The basic mistake is that, even if aged persons stay off the streets as much as young people, one will still obtain a wrong estimate unless people of intermediate ages stay off the streets to the same degree.

To see in a rough way why this is so, think about the extreme case in which the aged and the young people never come on the streets. Then Paine's method would tell him there are no people 50 or over, which would certainly be false.

England's population. It may surprise you that Paine did not go to the English census office for the information he needed. He couldn't, because there was no real census in England until 1801. It gave about 9,000,000 as the population for England and Wales, but it provided no information about ages. Not until the census of 1821 were comprehensive figures by age available.

Of course Paine did use an estimate of the total population, presumably of England and Wales, although he does not make clear if Wales is included or if Scotland is included. Paine's figure of 420,000 for those 50 or older was obtained by dividing 16 into the total population. Call the total x, so that

$$\frac{x}{16} = 420{,}000.$$

Hence, x must have been about 6,720,000, as you may have computed in Exercise 1. A few pages before the quotation given above, Paine used 7,000,000 as the population figure.

Exercise

5. There may be a rounding effect that explains the apparent inconsistency. Suppose that Paine decided to round his estimated number of elderly people to the closest 10,000. What range of population figures would then give rise to his 420,000? Is 7,000,000 in that range?

Paine does not explain where his population figure came from, but there were several population estimates that had been made not too long before Paine did his writing. In fact, there was much concern in England about population. Some writers thought that the population was decreasing at an alarming rate, in part because of emigration to the British colonies.

On the other hand, a few thinkers were reflecting on the possibility of overpopulation. The most renowned of these was Thomas Malthus, who published in 1798 the first edition of his famous book An Essay on Population. In it he gave the "population of the Island... [as] about seven millions". Malthus' major theme was how population growth would outstrip the supply of food, and some of his discussion resembles current writings about our contemporary population problems.

There was much controversy and many different estimates of population. In terms of the population of

England and Wales, some of these estimates were the following:

5,000,000 or less. Richard Price, 1777.

8,691,597. John Howlett, 1780. Howlett increases this to "very little less than nine millions" if dockyards, hospitals, work-houses, etc. are added.

8,447,200. George Chalmers, 1793.

11,000,000. Frederick Eden, 1800.

The 1801 census gave the population of England and Wales as 8,872,980. After a rough correction for the armed forces and other special groups, the number was raised to 9,168,000.

The proportion 50 or over. If it was hard enough to estimate the size of the population in Paine's time, it was harder still to estimate accurately the proportion of the population aged 50 or more.

There are few reliable contemporary records to help us, but one survey is thought to be relatively accurate and is worth mention. In the late eighteenth century there lived in the city of Carlisle, an English town near the Scottish border, a physician named John Heysham. Heysham was a prominent citizen of Carlisle, and a man of wide interests, including a concern with public health. He made careful surveys of the citizens of Carlisle, their ages, births, sicknesses, and deaths; and his summary records have been published. In fact, Heysham's death rates formed valuable basic data for early life insurance in England.

In 1780, Heysham found that the total population of Carlisle, including its surrounding villages, was 7,677 and of that population, 1,287 were 50 years old or older. That works out to a proportion of about 17%, or 1/6, 50 or older. Another survey, in 1788, gave the same result. On the other hand, Paine's figure of 1/16 corresponds to only 6 1/4%. That is substantially lower than 17%, and the difference is larger than one might expect between the streets of London and the quieter Carlisle area. We cannot be certain, but it seems plausible that Paine's estimate is much too low simply because older people spent less time in the street than others.

Some confirmation of Heysham's 17% comes from data collected in Sweden and Finland at about the same time. The astronomer H. Nicander was in charge of official statistics, and his results on the age distribution for 20 years ending in 1795 show that about 16% of the population was 50 or over. Thus Nicander's 16% and Heysham's 17% are both considerably larger than Paine's 6%.

Still further evidence is available from Denmark, Iceland, France, and perhaps other European countries. The percentages of people 50 or older around 1800 ranged from 13% to 20%, all much larger than Paine's 6%. In fact, the only exception seems to be the United States of America; there the first census with detailed age information (for white males only) was held in 1830. It showed that only 8% were 50 or older. This is hardly surprising because of the immigration into America of younger men hoping to settle in the rapidly expanding continent. We might wonder whether Paine's observations were really made in the streets of New York, not London.

In the British census of 1821, 14% of the population was 50 or older. Some later British demographers calculated backwards, using plausible assumptions, and estimated that 16% of the population in 1801 was 50 or older.

Thus data from well before 1792 to well after suggest that Paine's 6% figure was only about 1/3 of the correct figure. This difference is substantial in terms of Paine's purpose. He estimated how much financial aid an elderly person might need from the state, and he multiplied that by his estimated number of elderly people. Had he used 17% instead of 6% the total cost would have been almost three times as much, and it would have been easier to criticize Paine's proposed old-age assistance as impractical.

We see similar problems today. For example, many programs of government social aid turn out to cost more than initial estimates. One reason is that it is harder to count people who are in economic trouble, or outside the mainstream of the nation's life, than it is to count other citizens. In some economically depressed areas of our cities today, it is estimated that as much as 12% of the population is missed by the census.

Project

In this project, you are asked to plan and carry out a survey in a manner something like Paine's.

The survey is to find out what proportion of cars in your city or neighborhood is European- or Japanese-made rather than American made. First you must plan the survey carefully. Where should you make it and at what time of day? Pick a spot where you can stand safely and quietly while making your counts. Should you do your survey during morning or evening rush hours or in midmorning or midafternoon? Who should do the survey? Boys from the motor-mechanics class who may know cars very well, or anybody? If the latter, and the survey is made by several different people at different places and different times of day, should the class try to average the results somehow?

Before you make your actual survey, you should make some trial runs and see what difficulties you will meet. You should prepare form sheets to record the results of your surveys. You should decide whether to work alone or in teams of two, so that one can spot and the other record the counts.

Finally, after your class has made several surveys, you should compare results and talk over possible sources of error. Discuss whether the results of such surveys made in New York, Detroit, St. Louis, San Francisco and Quebec might be different, and if so, why. Try to get information to compare your results with the number of foreign car imports versus the number of American-made cars sold.

A second project might be to make a survey of the proportion of station wagons to other cars, or trucks to passenger cars.

A third project might be to determine the proportion of students to adults among the people who use your local public library.

Projects like these will give you some idea of the problems and difficulties that have to be solved if surveys are to be accurate and useful. Usually surveys have to be samples rather than complete censuses, because it may be too expensive or even impossible to do the latter.

We have not proposed projects parallel to Paine's survey because age cannot be told reliably by looking at people. In addition, sampling people "in the streets" is especially difficult.

Fruit Flies

RICHARD F. LINK

INTRODUCTION

Many times observations are made on characteristics that may be classified as falling into one of two classes. For example, if we toss a coin, the top face will be a head or a tail. In football, when a passer throws a pass, it will be complete or incomplete. A newborn baby will be a boy or a girl.

Sometimes we may arbitrarily classify data into two groups, even though a more refined measurement might be made. For example, we may measure the depth of a river, but for many purposes we are content to classify the river as being flooded or not flooded. A car has a continuously variable speed, but often it is sufficient to classify the speed as being above the speed limit or below. Problems dealing with two classes are often called binomial because of the "two names" of the classes.

If we assign the value "0" to one classification and the value "1" to the other classification, then we can quickly establish how many items in a group have a given classification. If, for example, ten cars are observed, we may assign a "0" to each car going below the speed limit and a "1" to each car going above the speed limit. If we now add the numbers assigned to the ten cars, we know how many are speeding. If all the cars receive a 0 (i.e., all are going below the speed

The author is at Artronic Information Systems, Inc., New York, New York, and Princeton University, Princeton, New Jersey.

limit), the sum is 0 but if three cars are above the speed limit, the sum is 3 since we would have assigned seven 0's and three 1's.

An important case arises when the numbers of items falling into each class are equal on the average. That is, the item being observed has a fifty-fifty chance of being in either category. Thus in tossing a fair coin the possibility of getting a head is fifty-fifty and in a long series of tosses of 100 such coins the numbers of heads and tails will each be on the average about equal to fifty. We shall discuss the fifty-fifty chance in some detail so that we may see how samples behave for this special case. But instead of coins we shall use a more interesting example from the field of genetics.

Genetics is a science that studies the inheritance of characteristics. Many types of plants and animals have been studied with the result that new types of grain, turkeys, etc., have been developed. These developments have been made possible by understanding how various traits are inherited. Scientists have found that characteristics that are inherited are determined by small particles which have been named genes. These particles are located in groups that form bodies called chromosomes which always appear in pairs.

In animals, one special pair of chromosomes determines the individual's sex. An animal with two chromosomes, called X chromosomes, is a female. If on the other hand the animal has one X chromosome and one Y chromosome, it is a male. The names "X" and "Y" are arbitrary and are conventionally used. During the process of reproduction, all chromosome pairs split into single units and offspring are produced with a random pairing of one chromosome of each type from each parent so that the chances are about fifty-fifty that the offspring gets two X chromosomes and is a female. The chances are also fifty-fifty that the offspring gets one X and one Y chromosome and is a male.

Most inherited traits are the result of the influences of many genes, but some traits are influenced by a single gene. Sometimes genes get changed; then so may the physical characteristics whose inheritance they control. Such a change in a gene is called a mutation. In the fruit fly there is a gene that controls eye size, and the gene has two variants. One is called the normal form, and the other called bar, because, if it is present, the eye of the fly becomes smaller. This gene is on the X chromosome, and in the female fly, three types of eyes can be distinguished, depending on whether the pair of chromosomes has two normal genes, one normal and one bar, or two bar genes. The male fly has only one X chromosome, and its eye is either normal or very small. The sketch below illustrates the various kinds of flies' eyes that can result.

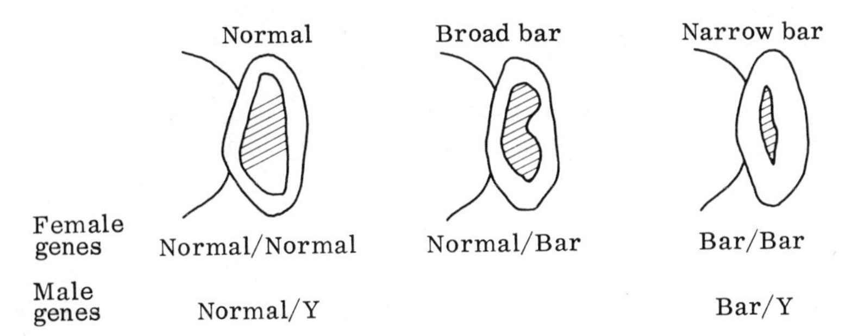

Fig. 1. Eye types of fruit flies exhibiting the bar characteristic.

Geneticists can conduct an experiment where they mate a female fly with broad bar eyes (indicating that she has one normal and one bar gene) with a male fly whose eye size is normal (indicating that it has a normal gene on its single X chromosome). The eyes of the offspring in this experiment will be either normal eyes or eyes with the bar characteristic, which in this instance will be shown by broad bar eyes in the females and narrow bar eyes in the males. The genetic characteristics of each parent mix randomly, and the following table shows why the chance that a fly's eye will be normal from this mating is 1/2 and, of course, the chance that it will exhibit a bar characteristic is also 1/2.

Female parent genes	Male parent genes: Normal	Male parent genes: Y
Normal	Normal female	Normal male
Bar	Broad bar female	Narrow bar male

Offspring in the top row will have normal eyes while those in the bottom row will have the bar characteristic. Each cell combination is equally likely because of the random mixing of the genes from each parent, and there are two cells in each row.

Suppose now that we have made such a cross and have examined two offspring chosen at random. We will assign a 0 to flies with normal eyes and a 1 to flies which have the bar characteristic, thus in this experiment we will observe a number which will be 0 if both flies are normal, 1 if one fly is normal and one barred, and 2 if both flies are barred. If we repeat this experiment a hundred times what pattern might we see for the numbers, i.e., how many 0's, 1's, and 2's might there be?

We saw above that the chance of an individual fly's eye being normal is 0.5. So as we examine our two offspring, the chance that the first one is normal is 0.5 and the chance that the second one is also normal is also 0.5. Let's call this the NN case. There are three other cases which are just as likely as the first, namely the NB, BN, and BB cases. Since the chances for any one of these cases are the same, they must each be 1/4. Now since the chance of getting two normal-eyed flies is 1/4, if we perform the experiment 100 times we should get NN about (1/4)100 or 25 times. Similarly we should get BB about 25 times and one B and one N about 50 times, because it includes both the BN and the NB cases.

Since we can't all do experiments with fruit flies, we may simulate such experiments. This means that we do an experiment which although physically different and easier to do, has the same mathematical characteristics. Thus to simulate the experiment above we might throw 2 pennies, count and tally the number of heads that appear, repeat this 100 times and see how many 0's, 1's, and 2's appear and are tallied. We actually did this 5 times and got the following results in such a simulation.

Number of heads	Simulation I	II	III	IV	V	Total
0	20	27	19	23	18	107
1	57	52	62	51	49	271
2	23	21	19	26	33	122

Certainly these results do not follow exactly the pattern 25, 50, 25, but then nobody should expect them to since chances are not certainties.

What pattern would we see if we examined 3 offspring instead of 2 and counted the number who had normal eyes? The count would be 0, 1, 2, or 3. Five simulations of this experiment gave the following table.

Number of heads	Simulation I	II	III	IV	V	Total
0	14	11	12	15	15	67
1	43	37	36	26	35	177
2	28	39	33	43	38	181
3	15	13	19	16	12	75

Again there is quite a variation in the results from experiment to experiment but some pattern is visible. The numbers 1 and 2 appear much more often than do 0 and 3. Theoretically they should appear just three times as often and theoretically 0 and 3 should appear the same number of times. Just as in the case of 2 flies where theory says 0 should appear $(1/2)^2 100$ or 25 times, so theory in this case says 0 should appear $(1/2)^3 100$, or about 12.5 times in 100 cases, and we note that our results are not far from that.

What is the general pattern? In the case of two flies it is:

Number of bars	0	1	2
Expected count	$1(\frac{1}{2})^2 \cdot 100$	$2(\frac{1}{2})^2 \cdot 100$	$1(\frac{1}{2})^2 \cdot 100$

In the case of three flies it is:

Number of bars	0	1	2	3
Expected count	$1(\frac{1}{2})^3 \cdot 100$	$3(\frac{1}{2})^3 \cdot 100$	$3(\frac{1}{2})^3 \cdot 100$	$1(\frac{1}{2})^3 \cdot 100$

Some of you may recognize the coefficients as having the pattern of the coefficients in

$$(a+b)^2 = 1 \cdot a^2 + 2 \cdot ab + 1 \cdot b^2$$

and

$$(a+b)^3 = 1 \cdot a^3 + 3 \cdot a^2 b + 3 \cdot ab^2 + 1 \cdot b^3.$$

Since we are looking here at the expansion of the binomial a+b, we call these coefficients the _binomial coefficients_ for the power 2 or 3 as the case may be.

The same pattern of the appearance of the appropriate binomial coefficients will be followed, at least approximately, if we examine four flies in each case, or five, or ten, or as many as we care to. For this reason this kind of an experiment is called a _binomial experiment_, and since the chance of a success on each trial is fifty-fifty, it is called an equilikely binomial experiment.

To see how the pattern develops, let's suppose that the pattern 1, 3, 3, 1 holds for 3 flies. What is the pattern for 4? Each of the counts 0, 1, 2, 3 could either be unchanged by the addition of a fourth fly, or be increased by 1. The chances are fifty-fifty for the 2 possibilities. Also the 16 possibilities could be represented as follows:

		New total				
		0	1	2	3	4
If new fly adds	1		1	3	3	1
	0	1	3	3	1	
New distribution		1	4	6	4	1

Similarly, adding a fifth fly would give

		New total					
		0	1	2	3	4	5
New fly adds	1		1	4	6	4	1
	0	1	4	6	4	1	
New distribution		1	5	10	10	5	1

In this way we can, in principle, build up any coefficient we want.

It can be shown that these binomial coefficients can be represented as quotients of factorials. Let N be the number in the sample, and R be the number of barred flies. When N = 5, the coefficient for R = 2 is

$$\frac{5!}{2!3!}.$$

where $X! = X(X-1) \ldots 2 \cdot 1$, and X is an integer greater than 1, and $1! = 0! = 1$.

And so the coefficient for N = 10, R = 5 is

$$\frac{10!}{5!5!} = \frac{10\times9\times8\times7\times6\times5\times4\times3\times2\times1}{5\times4\times3\times2\times1\times5\times4\times3\times2\times1} = \frac{10\times9\times8\times7\times6}{5\times4\times3\times2\times1} = 252.$$

The total of the coefficients for a given N is 2^N, as you can see from their formation, for example, $1 + 2 + 1 = 2^2$, $1 + 3 + 3 + 1 = 2^3$, etc. Consequently, the probability in the fifty-fifty case for N = 10, R = 5 is 252/1024 or about 0.25.

Let's now consider one more simulation when we look at 10 flies.

Table 1 shows that the chance of getting the sum of the 1's to be either "0" or "10" is quite small. One such case occurred in this simulation. Theoretically, this should happen about one time in 500. The value 5 is the most frequently occurring value of "X", as you might expect. It occurred one time in four, or

Table 1. Simulation of number of offspring with bar characteristic resulting from cross of broad bar female with a normal male

Number of flies with barred eyes in sample of 10	Tally	Frequency
10	1	1
9		0
8	111	3
7	~~1111~~ ~~1111~~ 111	13
6	~~1111~~ ~~1111~~ ~~1111~~	15
5	~~1111~~ ~~1111~~ ~~1111~~ ~~1111~~ ~~1111~~	25
4	~~1111~~ ~~1111~~ ~~1111~~ ~~1111~~ 11	22
3	~~1111~~ ~~1111~~ 1111	14
2	1111	4
1	111	3
0		0
		100

25 times out of 100 tries. Theoretically, it should occur with about that frequency.

Exercises

1. In the experiment described in the text, suppose that instead of looking at 10 flies, we had looked at only 2 flies for each cross. What fraction of crosses would yield 2 normal flies, 1 bar and 1 normal, and 2 bar flies? Why don't these three outcomes occur with equal frequency?

2. What would be the various outcomes and frequencies of these outcomes if we look at 6 flies?

3. Suppose we were to make many crosses of a bar male with a normal female. In what fraction of the crosses will the bar characteristic occur in the male offspring.

4. Events whose chance or frequency of occurrence is only around 10% may be considered "rare". If R or fewer successes in 10 trials are rare, how big can R be? Use Table 1.

5. Toss two coins 40 times, tossing each one separately. Score two tails 0, one tail and one head 1, and two heads 2. Construct the frequency distribution for this score. Does this appear to be close to the distribution you calculated in Exercise 1?

6. If N = 10, R = 1, evaluate the binomial coefficient.

7. If N = 10, R = 0, evaluate the binomial coefficient.

8. Show that N = 10, R = 0 gives the same binomial coefficient as N = 10, R = 10.

9. Show that the binomial coefficient for N = 10, R = 2 is the same as that for N = 10, R = 8.

10. Explain why the binomial coefficient for N and R is the same as that for N and N-R.

11. Use the result of Exercise 8 to find exactly the probability of 0 or 10 bar-eyed fruit flies in our N = 10 experiment, and thus verify that the chances are about 1 in 500.

12. Find the chance of getting 4 or 6 bar-eyed fruit flies in the N = 10 experiment and compare the theoretical probability with that observed in the simulation of Table 1.

Index

Index

73